THE HISTORY OF BRITISH GEOLOGY

A Bibliographical Study

THE HISTORY OF BRITISH GEOLOGY

A Bibliographical Study

John Challinor

Sometime Senior Lecturer in the Department of Geology,
University College of Wales, Aberystwyth

DAVID & CHARLES : NEWTON ABBOT

ISBN O 7153 5343 8

COPYRIGHT NOTICE

© John Challinor 1971

Set in ten point Baskerville
and printed in Great Britain
by Bristol Typesetting Company Limited
for David & Charles (Publishers) Limited
South Devon House Newton Abbot Devon

Contents

Preface

The purpose of this book is to provide a short introduction to a great and hitherto entirely neglected subject: the general history of British geology. The method adopted is to review the written records of the progress of knowledge.

The first main section consists of a selected list of works arranged chronologically, year by year. The list includes works of purely original research, works of summary and generalisation, expositions of principles, and various combinations of these. It is hoped that no work of the first importance has been left out, but the list is necessarily a selection from the vast store of records which constitute the science of geology as developed in Britain. As geology is a *science*, its history is a part of itself.

In the second main section these works are treated in a thematic context; a series of short essays, summaries, and compilations arranged in an order which seems most reasonably to portray the evolving story.

The three appendixes list (A) additional works and works of commentary referred to in Section II, (B) authors with short biographical statements, and (C) names of places, stratigraphical divisions and fossils, and refers to Sections I and II. All the parts of the work are connected together by means of cross-references.

The parts of the book which deal with the progress of British geology up to the year 1822 are based on the present writer's articles in the *Annals of Science* for 1953 and 1954, and for 1970. Many further details and extracts will be found in those papers. Historical allusions, particularly to British geology, form the basis of his *Dictionary of geology*.

To attempt to set this sketch against the background of the general progress of geology in Europe and the world would be beyond our scope. General histories are provided in their several ways by such well-known works as Zittel's *History of geology and palaeontology* (English version, 1901), Geikie's *Founders of geology* (1905 edition), Woodward's short *History of geology* (1911), and Adams's *Birth and development of the geological sciences* (1938).

I

The Primary Literature:
a chronological listing

NOTES

1 The items are arranged under each year in alphabetical order of authors. Each is given a serial number so that references may be made in Section II and Appendixes B and C.

2 Abbreviations of the titles of periodicals are given in a generally accepted form except for the following which are frequently cited : *TGS*—Transactions of the Geological Society of London; *QJGS*— Quarterly Journal of the Geological Society of London.

3 Where there are several editions of a work, that work is listed under the year of its first appearance; the existence of significant later editions being indicated by 'L'. In some cases a particular edition is specified.

4 Items which are published as separate parts or articles, but which are so closely connected as to be parts of the same subject, are here placed under one heading in the year of their first appearance, the span of years being given at the end of the entry. Items which are not so closely connected but which nevertheless may be considered as part of a series are each placed separately under the appropriate year, the serial numbers of the other related items being referred to in brackets.

5 The numbers in italics at the ends of the entries refer to the themes in Section II.

6 Numbers followed by a colon indicate volume numbers.

1538 (circa)
[1] Leland, J. MS. *The itinerary of John Leland . . . begunne about 1538*. London, 1710. Bodleian Library, Oxford. L.   *1*

1586
[2] Camden, W. *Britannia*. London. First English edition, by P. Holland, London, 1610. L; edited by E. Gibson, 1695.   *1*

1603
[3] Owen, G. *Description of Pembrokeshire*. MS edited by R. Fenton, *Cambrian Register* for 1796 (1799); and, with introduction and notes, by H. Owen, *Cymmrodorion Record Series*, no. 1, 1892.   2, 7

1661
[4] Childrey, J. *Britannica Baconica*. London.   3

1665
[5] Hooke, R. *Micrographia*. London.   3

1669
[6] Somner, W. *Chartham news: or a brief relation of some strange bones there lately digged up*. London. (also *Phil. Trans. Roy. Soc.*, 22 : 882-93, 1701.)   3

1671
[7] Lister, M. Concerning petrify'd shells. *Phil. Trans. Roy. Soc.*, 5 : 2281-4.   3

1672
[8] Sinclair, G. A short history of coal. in *The hydrostaticks*. Edinburgh.   4, 8

1676
[9] Beaumont, J. Concerning rock-plants [crinoids] growing in the lead mines of Mendip Hills. *Phil. Trans. Roy. Soc.*, 11 : 724-42; 13 : 276-80, 1676-83.   3

1677
[10] Plot, R. *The natural history of Oxfordshire*. Oxford. L.   3

1678
[11] Lister, M. De lapidibus ejusdem insulae ad cochlearum quandam imaginem figuratis. in *Historia animalium Angliae*. London.   3

1684
[12] Lister, M. An ingenious proposal for a new sort of maps of countrys. *Phil. Trans. Roy. Soc.*, 14 : 739-46.   7

1686
[13] Plot, R. *The natural history of Staffordshire*. Oxford.   *3, 4*

1688
[14] Lister, M. De conchites. in *Historia sive synopsis methodica conchyliorum*. London.   *3*

1695
[15] Lhwyd, E. [Observations.] in E. Gibson's edition of W. Camden's *Britannia*. London.   *3, 5*
[16] Woodward, J. *An essay towards a natural history of the Earth*. London. L.   *4*

1699
[17] Lhwyd, E. *Lithophylacii Britannici Ichnographia*. London. L; 2nd, 1760.   *3, 32*

1700
[18] Brewer, J. Concerning beds of oyster-shells found near Reading. *Phil. Trans. Roy. Soc.*, 22 : 484-6.   *4*
[19] Leigh, C. *The natural history of Lancashire, Cheshire and the Peak in Derbyshire*. Oxford.   *3, 4, 5*

1705
[20] Hooke, R. A discourse of earthquakes. in *The posthumous works of Robert Hooke*, edited by R. Waller. London.   *3*

1709
[21] Robinson, T. *An essay towards a natural history of Westmorland and Cumberland*. London.   *4*

1712
[22] Morton, J. *The natural history of Northamptonshire*. London.   *3*

1719
[23] Strachey, J. A curious description of the strata observ'd in the coal-mines of Mendip in Somersetshire, *and* An account of the strata in coal-mines, etc. *Phil. Trans. Roy. Soc.*, 30 : 968-73; 33 : 395-8. Together as the pamphlet *Observations on the different strata of earths and minerals, more particularly of such as are found in the coal-mines of Great Britain*. 1719-27.   *4, 6, 13, 17*

[24] Stukeley, W. An account of the impression of the almost entire sceleton of a large animal in a very hard stone. *Phil. Trans. Roy. Soc.,* 30 : 963-8.   *3, 32*

1723
[25] Holloway, B. An account of the pits of fullers-earth in Bedford-shire. *Phil. Trans. Roy. Soc.,* 32 : 419-21.   7

1728
[26] Woodward, J. *Fossils of all kinds digested into a method.* London.   3
[27] Woodward, J. *An attempt towards a natural history of the fossils of England.* [A catalogue.] London. 1728-9.   3

1730
[28] Dale, S. *The history and antiquities of Harwich and Dovecourt, first collected by Silas Taylor, and now much enlarged, with notes and observations relating to natural history.* London.   3

1743
[29] Packe, C. *A new philosophico-chorographical chart of East Kent. Accompanied by an explanation.* London.   5

1754
[30] Owen, E. *Observations on the earths, rocks, stones and minerals for some miles about Bristol.* London.   5

1758
[31] Borlase, W. *The natural history of Cornwall.* Oxford.   5
[32] Chapman, W. and Wooller, J. Fossil skeleton found near Whitby. *Phil. Trans. Roy. Soc.,* 50 : 688-91, 786-90.   *3, 32*

1760
[33] Michell, J. Conjectures concerning the cause, and observations upon the phaenomena of earthquakes. *Phil. Trans. Roy. Soc.,* 51 : 566-634.   *4*

1766
[34] Solander, D.C. [Figures and descriptions of Tertiary fossils.] in G. Brander and D. C. Solander, *Fossilia Hantoniensia.* London.   3

1774
[35] Banks, J. [Account of Staffa.] in T. Pennant, *A tour in Scotland and voyage to the Hebrides in 1772.* London.   5

1778

[36] Pryce, W. *Mineralogia Cornubiensis.* London.   5
[37] Whitehurst, J. General observations on the strata in Derby-
    shire. in *An inquiry into the original state and formation of
    the Earth.* London. L.   *4, 8, 17*

1779

[38] Walcott, J. *Descriptions and figures of petrifactions found in
    the quarries, gravel pits, etc. near Bath.* Bath.   *3*

1782

[39] Strange, J. Antiquities in Monmouthshire and Glamorganshire.
    Archaeologia, 6 : 6-38.   7

1788

[40] Hutton, J. Theory of the Earth, or an investigation of the
    laws observable in the composition, dissolution, and
    restoration of land upon the globe. *Trans. Roy. Soc. Edinb.*,
    1 : 209-304.   *8, 9, 10, 12, 16*
[41] Michell, J. [List of strata.] MS described by J. Farey. in Copy
    of a list of the principal British strata by the late Rev. John
    Michel, communicated by Mr. John Farey to Mr. Tilloch.
    Phil. Mag. 36 (1810): 102.   *6, 17*
[42] Michell, J. and Cavendish, H. [Letter on strata to Henry
    Cavendish and Cavendish's reply.] in A. Geikie. *Memoir
    of John Michell*, (1918). Cambridge.   7

1789

[43] Williams, J. *The natural history of the mineral kingdom.*
    Edinburgh. L.   *4, 10, 11, 15, 16, 17*

1790

[44] Hall, J. and Hutton, J. Remarks and observations on granite.
    Trans. Roy. Soc. Edinb., 3 (1794): part 1, 8-12; part 2,
    77-85.   *12*

1791

[45] Smeaton, J. *A narrative of the building and a description of the
    construction of the Eddystone lighthouse.* London.   7

1794

[46] Smith, W. [Geological sections.] MS described by J. Phillips,
    Memoirs of William Smith (1844), 59; and by L. R. Cox,
    Proc. Yorks. Geol. Soc., 25 (1942) : 8-9. 1794-1809.   *19*

1795

[47] Hutton, J. *Theory of the Earth, with proofs and illustrations.* Edinburgh. (An additional volume, edited with preface and notes by A. Geikie, 1899. London, Geol. Soc.)

9, 11, 12, 13, 14, 15

1797

[48] Faujas Saint Fond, B. *Voyage en Angleterre, en Écosse et aux Îles Hébrides.* Paris. L; English edition, London, 1799; English edition, with notes and a memoir, by A. Geikie. Glasgow, 1907.   *8, 11, 15*

[49] Maton, W. G. *Observations on the western counties of England, made in the years 1794 and 1796.* Salisbury.   *16, 17*

1799

[50] Smith, W. [Geological map of the country round Bath.] MS described by J. W. Judd, *Geol. Mag.* 34 (1897): 443; and by T. Sheppard, *Proc. Yorks. Geol. Soc.,* 19 (1917): 103, 109, pl. XI.   *19*

[51] Smith, W. [Table of strata.] MS described by W. H. Fitton, *Lond. and Edinb. Phil. Mag. Jnl Sci.,* 2 (1833): 46; and by J. Phillips. *Memoirs of William Smith.* 1844, 30.   *29*

1800

[52] Jameson, R. *Mineralogy of the Scottish Isles.* Edinburgh.

8, 11, 23

1801

[53] Smith, W. [Small geological maps of England and Wales.] MS described by L. R. Cox, *Proc. Yorks. Geol. Soc.,* 25 (1942): 26-30, pl. II. 1801-3.   *19*

1802

[54] Playfair, J. *Illustrations of the Huttonian Theory of the Earth.* Edinburgh.   *8, 9, 11, 13, 14, 17*

1804

[55] Parkinson, J. *Organic remains of a former world.* London. 1804-11. L.   *20, 28, 32*

[56] Sowerby, J. *British mineralogy.* London. 1804-17.   *20*

1805

[57] Jameson, R. *A mineralogical description of the County of Dumfries.* Edinburgh.   *19*

[58] Hall, J. [Experiments on igneous rocks and on the effects of pressure.] *Trans. Roy. Soc. Edinb.*, 5 : 43-75; 6 : 71-187; 7 : 79-108. 1805-15.   *21*

1807
[59] Farey, J. [Geological sections.] MS described by T. Ford. *Mercian Geologist*, 2 (1967): 41-9. 1807-8.   *19*

1808
[60] Necker, L. A. *Scotland coloured according to the rock formations* [map]. Facsimile reproduction of MS. Edinburgh Geological Society, 1939.   *23*

1809
[61] Forster, W. *A treatise on a section of the strata, from Newcastle-upon-Tyne to the mountain of Cross Fell*. Alston. L; edited by W. Nall, with a memoir of the author. 1883.   *25*
[62] Martin, W. *Petrificata Derbiensia; or figures and descriptions of petrifactions collected in Derbyshire*. Wigan. (One part published at Wigan, 1793).   *20*
[63] Martin, W. *Outlines of an attempt to establish a knowledge of extraneous fossils on scientific principles*. Macclesfield.   *20*

1811
[64] Aikin, A. Observations on the Wrekin and the great coal-field of Shropshire. *TGS*, 1 : 191-212.   *24*
[65] Farey, J. *General view of the agriculture and minerals of Derbyshire*. Vol. 1. London.   *25, 29, 30*
[66] Holland, H. A sketch of the natural history of the Cheshire rock-salt district. *TGS*, 1 : 38-62.   *24*
[67] Horner, L. On the mineralogy of the Malvern Hills. *TGS*, 1 : 281-321.   *24, 46*
[68] Parkinson, J. Observations on some of the strata in the neighbourhood of London, and on the fossil remains contained in them. *TGS*, 1 : 324-54.   *24, 28*
[69] Watson, W. *A delineation* [section] *of the strata of Derbyshire, forming the surface from Bolsover in the east to Buxton in the west*. Sheffield. (See also 73.)   *25*

1812
[70] Sowerby, J. and Sowerby, J. de C. *The mineral conchology of Great Britain*. London. 1812-26, with an additional unfinished volume, 1846.   *20*

B

1813

[71] Bakewell, R. *An introduction to geology.* London. L.   22

[72] Townsend, J. *The character of Moses established for veracity as an historian: recording events from the Creation to the Deluge.* Bath and London.   22, 29

[73] Watson, W. *A section of the strata forming the surface in the vicinity of Matlock Bath in Derbyshire.* Chesterfield. (See also 69.)   25

1814

[74] Berger, J. F. Mineralogical account of the Isle of Man. *TGS*, 2 : 29-59.   24

[75] Webster, T. On the freshwater formations of the Isle of Wight, with some observations on the strata over the Chalk in the south-east part of England. *TGS*, 2 : 161-254.   24, 26

1815

[76) Smith, W. *A delineation* [map] *of the strata of England and Wales, with part of Scotland.* (Scale, 5 miles to 1 inch. Together with *A memoir to the map*). London.   27, 28, 29

1816

[77] Horner, L. Sketch of the geology of the south-western part of Somersetshire. *TGS*, 3 : 338-84.   24

[78] Phillips, W. An outline of the geology of England and Wales. in *Outlines of mineralogy and geology*, 2nd edition. London.   22

[79] Smith, W. *Strata identified by organized fossils.* London. 1816-19.   28

[80] Webster, T. Observations on the strata of the Isle of Wight and their continuation in the adjacent parts of Dorsetshire. in H. C. Englefield. *A description of the principal picturesque beauties, antiquities, and geological phenomena of the Isle of Wight.* London, Payne and Foss.   26, 30

1817

[81] Bright, R. On the strata in the neighbourhood of Bristol. *TGS*, 4 : 193-215.   24

[82] Buckland, W. Description of an insulated group of rocks of slate and greenstone on the east side of Appleby. *TGS*, 4 : 105-16.   24

[83] Buckland, W. The Plastic Clay formation. *TGS*, 4 : 277-304.   24

[84] Hennah, R. The limestone of Plymouth. *TGS*, 4 : 410-12; 5 (part 2) : 619-24, and *A succinct account of the lime rocks of Plymouth*. Plymouth, Curtis, 1825. 1817-25.   24
[85] Macculloch, J. On the parallel roads of Glen Roy. *TGS*, 4 : 314-92.   24
[86] Smith, W. *Stratigraphical system of organized fossils*. London.   28, 29
[87] Smith, W. [Charts of geological sections.] Some lithographed, and others printed and published in London. 1817-24.   27
[88] Winch, N. Observations on the geology of Northumberland and Durham. *TGS*, 4 : 1-101.   24

1818
[89] Buckland, W. Order of superposition of strata in the British Islands. in W. Phillips, *A selection of facts* [etc.] London.   29, 45
[90] Phillips, W. *A selection of facts from the best authorities, arranged so as to form an outline of the geology of England and Wales*. London.   22

1819
[91] Greenough, G. B. *A geological map of England and Wales*. (Scale, 6 miles to 1 inch.) London, Geol. Soc.   27, 29
[92] Greenough, G. B. *A critical examination of the first principles of geology*. London, Longman.   22
[93] Macculloch, J. *A description of the western islands of Scotland including the Isle of Man*. London, Constable.   23
[94] Phillips, W. Remarks on the Chalk cliffs in the neighbourhood of Dover, and on the Blue Marle [Gault] covering the Green Sand, near Folkestone. *TGS*, 5 (part 1) : 16-51.   24
[95] Smith, W. *A new geological atlas of England and Wales*. London. 1819-24.   27

1820
[96] Boue, A. *Essai géologique sur l'Écosse*. Paris.   23
[97] Smith, W. *A new geological map of England and Wales*. (Scale, 15 miles to 1 inch.) London.   27

1821
[98] Buckland, W. Description of the quartz rock of the Lickey Hill in Worcestershire, and of the strata immediately surrounding it. *TGS*, 5 (part 2) : 506-44.   24
[99] Conybeare, W. D. and De la Beche, H. T. Notice of the discovery of a new fossil animal [*Plesiosaurus*] forming a link

between the *Ichthyosaurus* and crocodile, together with general remarks on the osteology of the *Ichthyosaurus*. *TGS*, 5 (part 2) : 558-94. (See also 102 and 113.)   *32*

[100] Henslow, J. S. Suplementary observations to Dr. Berger's account of the Isle of Man. *TGS*, 5 (part 2) : 482-505.   *24*

[101] Miller, J. S. *A natural history of the Crinoidea*. Bristol.   *20*

1822

[102] Conybeare, W. D. Additional notices on the fossil genera *Ichthyosaurus* and *Plesiosaurus*. *TGS*, 2nd series, 1 (part 1) : 103-23. (See also 99 and 113.)   *32*

[103] Conybeare, W. D. and Phillips, W. *Outlines of the geology of England and Wales*. London, Phillips.   *22, 29*

[104] De la Beche, H. T. Remarks on the geology of the south coast of England, from Bridport Harbour, Dorset, to Babbacombe Bay, Devon. *TGS*, 2nd series, 1 (part 1) : 40-7. (See also 115, 116, 117, 122 and 130.)   *33*

[105] Henslow, J. S. Geological description of Anglesea. *Trans. Camb. Phil. Soc.*, 1 : 359-452.   *34*

[106] Mantell, G. *The fossils of the South Downs: or illustrations of the geology of Sussex*. London, Relfe.   *34*

[107] Parkinson, J. *Outlines of oryctology: an introduction to the study of fossil organic remains*. London. L.   *20*

[108] Young, G. *A geological survey of the Yorkshire coast*. Whitby, Clark.   *34*

1823

[109] Buckland, W. *Reliquiae diluvianae*. London, Murray. L.   *31*

[110] Otley, J. *A concise description of the English Lakes, and adjacent mountains, with general directions to tourists; and observations on the mineralogy and geology of the district*. Keswick, Otley. L.   *30*

1824

[111] Buckland, W. Notice on the *Megalosaurus* or great fossil lizard of Stonesfield. *TGS*, 2nd series, 1 (part 2) : 390-6.   *32*

[112] Buckland, W. and Conybeare, W. D. Observations on the south-western coal-district of England. *TGS*, 2nd series, 1 (part 2) : 210-316.   *34*

[113] Conybeare, W. D. On the discovery of an almost perfect skeleton of the *Plesiosaurus*. *TGS*, 2nd series, 1 (part 2) : 381-9. [See also 99 and 102.]   *32*

1825
[114] Mantell, G. A. On the teeth of the *Iguanodon*, a newly-discovered fossil herbivorous reptile. *Phil. Trans. Roy. Soc.*, 115 : 179-86.   *32*

1826
[115] De la Beche, H. T. On the geology of southern Pembrokeshire. *TGS*, 2nd series, 2 (part 1) : 1-20. [See also 104, 116, 117, 122 and 130.]   *33*
[116] De la Beche, H. T. On the Lias of the coast in the vicinity of Lyme Regis, Dorset. *TGS*, 2nd series, 2 (part 1) : 21-30. (See also 104, 115, 117, 122 and 130.)   *33*
[117] De la Beche, H. T. On the Chalk and sands beneath it (usually termed green-sand) in the vicinity of Lyme Regis, Dorset, and Beer, Devon. *TGS*, 2nd series, 2 (part 1) : 109-18. [See also 104, 115, 116, 122 and 130.]   *33*

1827
[118] Murchison, R. I. On the coal-field of Brora, in Sutherlandshire, and some other stratified deposits in the north of Scotland. *TGS*, 2nd series, 2 (part 2) : 293-326. (See also 120.)   *34*

1828
[119] Martin, P. J. *A geological memoir on part of western Sussex*. London.   *34*
[120] Murchison, R. I. Supplementary remarks on the strata of the Oolitic series in Sutherland, Ross, and the Hebrides. *TGS*, 2nd series, 2 (part 3) : 353-68. (See also 118.)   *34*

1829
[121] Buckland, W. On the discovery of a new species of pterodactyl in the Lias at Lyme Regis. *TGS*, 2nd series, 3 (part 1) : 217-22.   *32*
[122] De la Beche, H. T. On the geology of Tor and Babbacombe bays, Devon. *TGS*, 2nd series, 3 (part 1) : 161-70. (See also 104, 115, 116, 117 and 130.)   *33*
[123] Phillips, J. *Illustrations of the geology of Yorkshire: The Yorkshire Coast*. York, Wilson. L.   *33, 35*
[124] Sedgwick, A. On the geological relations and internal structure of the Magnesian Limestone and the lower portions of the New Red Sandstone series. *TGS*, 2nd series, 3 (part 1) : 37-124. (See also 136.)   *34*

1830

[125] Lyell, C. *Principles of geology*. London, Murray. 1830-33. L.
                                                                    35

1831

[126] De la Beche, H. T. *A geological manual*. London, Treuttel
      and Würtz. L.   *35*
[127] Sedgwick, A. Notices of oral accounts of the geology of North
      Wales. in *Proc. Geol. Soc. Lond; Minutes Camb. Phil.
      Soc;* and *Rep. Br. Ass. Advmt. Sci.* 1831-8.   *36*

1832

[128] Smith, S. Stratification in [Map of] Hackness Hills. (Scale,
      6 inches to 1 mile.) London.   *34*

1834

[129] Phillips, J. *A guide to geology*. London, Longman. L; 2nd,
      1835.   *35*

1835

[130] De la Beche, H. T. and Buckland, W. On the geology of the
      neighbourhood of Weymouth and the adjacent parts of
      the coast of Dorset, *TGS*, 2nd series, 4 (part 1) : 1-46. (See
      also 104, 115, 116, 117 and 122.)   *33*
[131] Sedgwick, A. Remarks on the structure of large mineral
      masses. *TGS*, 2nd series, 3 (part 3) : 461-86.   *41*
[132] Sedgwick, A. Introduction to the general structure of the
      Cumbrian mountains. *TGS*, 2nd series, 4 (part 1) : 47-68.
      (See also 149, and 153.)   *30*

1836

[133] Buckland, W. *Geology and mineralogy considered with
      reference to natural theology*. (Bridgewater Treatise.)
      London, Pickering. L.   *32, 39*
[134] Fitton, W. H. Observations on some of the strata between
      the Chalk and the Oxford Oolite, in the south-east of
      England. *TGS*, 2nd series, 4 (part 2) : 103-389.   *37, 44*
[135] Phillips, J. *Illustrations of the geology of Yorkshire: the
      Mountain Limestone district*. London, Murray.   *37, 44*
[136] Sedgwick, A. On the New Red Sandstone series in the basin
      of Eden and north-western coasts of Cumberland and
      Lancashire. *TGS*, 2nd series, 4 (part 2) : 383-407. (See also
      124.)   *34*

1837
[137] Phillips, J. *A treatise on geology.* London, Longman. 1837-9.
35

1838
[138] Lyell, C. *Elements of geology.* London, Murray. L; some
editions entitled *A manual of elementary geology.*   35

1839
[139] De la Beche, H. T. *Report on the geology of Cornwall, Devon,
and West Somerset.* London, Longman.   45
[140] Maclaren, C. *A sketch of the geology of Fife and the Lothians.*
Edinburgh, Black. L.   *34, 43*
[141] Murchison, R. I. *The Silurian system.* London, Murray.
36, 45, 46

1840
[142] Bowerbank, J. S. *The fossil fruits of the London Clay.*
London, Van Voorst.   39
[143] Prestwich, J. On the geology of Coalbrook Dale. *TGS,* 2nd
series, 5 (part 3): 413-95.   34
[144] Sedgwick, A., Murchison, R. I., and Lonsdale, W. On the
physical structure of Devonshire and on its older stratified
deposits and on the age of the limestones of South Devonshire. *TGS,* 2nd series, 5 (part 3): 633-704, 721-38.   45

1841
[145] Miller, H. *The Old Red Sandstone.* Edinburgh, Constable.
L.   45
[146] Phillips, J. *Figures and descriptions of the Palaeozoic fossils of
Cornwall, Devon, and West Somerset.* London, Longman.
45
[147] Ramsay, A. C. *The geology of the island of Arran.* Glasgow,
Griffin.   34

1842
[148] Godwin-Austen, R. A. C. On the geology of the south-east
of Devonshire. *TGS,* 2nd series, 6 (part 2): 433-89.   45
[149] Sedgwick, A. Letters on the geology of the Lake District. in
J. Hudson. *Complete guide to the Lakes.* Kendal, Hudson.
1842-53. [See also 132, and 153.]   30

1845
[150] Brodie, P. B. *A history of the fossil insects in the Secondary
rocks of England.* London, Van Voorst.   39

[151] Morris, J. *A catalogue of British fossils.* London, Morris. L.
                                                                    39

[152] Ramsay, A. C. [Sections across Central Wales.] *Geol. Surv.,
      Horizontal Sections,* particularly sheets 4, 5 and 6. London,
      Geol. Surv.   *51*

[153] Sedgwick, A. Papers on the geology of the Lake District.
      QJGS, 1 : 442-50; 2 : 106-31; 3 : 133-64; 4 : 216-25. 1845-
      8. (See also 132 and 149.)   *30*

                              1846
[154] De la Beche, H. T. On the formation of the rocks of South
      Wales and south-west England. in *Memoirs of the
      Geological Survey* vol. 1, 1-296.   *44*

                              1848
[155] Phillips, J. The Malvern Hills compared with the Palaeozoic
      district of Abberley, Woolhope, May Hill, Totworth, and
      Usk [with a] palaeontological appendix by J. Phillips and
      J. W. Salter. in *Memoirs of the Geological Survey,* vol. 2,
      pt 1.   *46*

[156] Wood, S. V., the elder. *The Crag Mollusca.* London,
      Palaeontogr. Soc., 1848-82.   *39*

                              1849
[157] Forbes, E. *Figures and descriptions of British organic remains.*
      Decades 1-4. London, Geol. Surv., 1849-52.   *39*

                              1850
[158] Jones, T. R. Monographs on Entomostraca : *Entomostraca of
      the Cretaceous formation of England,* 1850; (and Sherborn,
      C.D.) *Tertiary Entomostraca of England,* 1857-89; *Fossil
      Estheriae,* 1863; (and others) *British fossil bivalved Ento-
      mostraca from the Carboniferous formations,* 1874-84; (and
      Woodward, H.) *British Palaeozoic Phyllopoda,* 1888-99;
      (and Hinde, G. J.) *Supplementary monograph of the
      Cretaceous Entomostraca of England and Ireland,* 1890.
      London, Palaeontogr. Soc., 1850-99.   *39*

[159] Milne-Edwards, H. and Haime, J. *British fossil corals.* London,
      Palaeontogr. Soc., 1850-5.   *39*

[160] Prestwich, J. On the structure of the strata between the
      London Clay and the Chalk in the London and Hampshire
      Tertiary systems. *QJGS,* 6 : 252-81; 8 : 235-64; 10 : 75-
      170. 1850-4.   *42*

1851

[161] Davidson, T. *British fossil Brachiopoda*. London, Palaeontogr. Soc., 1851-86. *39*

[162] De la Beche, H. T. *The geological observer*. London, Longman. L. *38*

[163] Owen, R. *Mesozoic fossil reptiles*. London, Palaeontogr. Soc., 1851-89. *32, 39*

[164] Prestwich, J. *A geological inquiry respecting water-bearing strata of the country around London*. London, Van Voorst. *42*

[165] Sorby, H. C. On the microscopical structure of the Calcareous Grit of the Yorkshire coast. *QJGS*, 7 : 1-6. *48*

1852

[166] Ramsay, A. C. On the superficial accumulations and surface markings of North Wales. *QJGS*, 8 : 371-6. (See also 184.) *40*

1853

[167] Jukes, J. B. *The geology of the South Staffordshire coalfield*. London, Geol. Surv. L. *50*

[168] Sedgwick, A. On a proposed separation of the so-called Caradoc Sandstone into two distinct groups. *QJGS*, 9 : 215-30. *36*

1854

[169] Murchison, R. I. *Siluria*. London, Murray. L. *36*

[170] Salter, J. W. and Aveline, W. T. On the 'Caradoc Sandstone' of Shropshire. *QJGS*, 10 : 62-75. *36*

1855

[171] Pengelly, W. Observations on the geology of the south-western coast of Devonshire. *Trans. Roy. Geol. Soc. Cornwall*, 7 : 291-7. *45*

[172] Phillips, J. *Manual of geology*. London, Griffin. *38*

[173] Sedgwick, A. and M'Coy, F. *A synopsis of the classification of the British Palaeozoic rocks, with a systematic description of the fossils*. Cambridge, University Press. *36*

1856

[174] Forbes, E. *The Tertiary fluvio-marine formation of the Isle of Wight*. London, Geol. Surv. *26*

[175] Godwin-Austen, R. A. C. On the possible extension of the Coal Measures beneath the south-eastern part of England. *QJGS*, 12 : 38-73. *47*

1857

[176] Jukes, J. B. *The student's manual of geology.* Edinburgh, Black. L.   *38, 52*

[177] Nicol, J. On the red sandstone and conglomerate [Torridonian] and the super-posed quartz-rocks, limestones, and gneiss of the north-west coast of Scotland. *QJGS,* 13 : 17-39. (See also 187.)   *54*

[178] Wright, T. *The British fossil Echinodermata of the Oolitic formations.* London, Palaeontogr. Soc., 1857-80.   *39*

1858

[179] Jamieson, T. F. The Pleistocene deposits of Aberdeenshire. *QJGS,* 14 : 509-32. (See also 183, 191, 201, 216 and 361.)   *40*

[180] Sorby, H. C. On the microscopic structure of crystals, indicating the origin of minerals and rocks. *QJGS,* 14 : 453-500.   *48*

1859

[181] Darwin, C. *The origin of species.* London, Murray. L.   *64*

[182] Ramsay, A. C. Geological map of England and Wales. (Scale 12 miles to 1 inch.) London, Stanford. L.   *49*

1860

[183] Jamieson, T. F. The drift and rolled gravel of the north of Scotland. *QJGS,* 16 : 347-73. (See also 179, 191, 201, 216, and 361.)   *40*

[184] Ramsay, A. C. *The old glaciers of Switzerland and North Wales.* London, Longman. (See also 166.)   *40*

1861

[185] Geikie, A. and Howell, H. H. *The geology of the neighbourhood of Edinburgh.* London, Geol. Surv. L; revisions in 1910 and 1962.   *43*

[186] Geikie, A. On the chronology of the trap rocks of Scotland. *Trans. Roy. Soc. Edinb.,* 22 : 633-53.   *43*

[187] Nicol, J. On the structure of the north-west Highlands and the relations of the gneiss, red sandstone [Torridonian], and quartzite of Sutherland and Ross-shire. *QJGS,* 17 : 85-113. (See also 177.)   *54*

1862

[188] Bristow, H. W. *The geology of the Isle of Wight.* London, Geol. Surv. L.   *26*

[189] Pengelly, W. and Heer, O. The lignites and clays of Bovey Tracey. *Phil. Trans. Roy. Soc.*, 152 : 1019-86.   *50*

1863

[190] Geikie, A. On the phenomena of the glacial drift of Scotland. *Trans. Geol. Soc. Glasg.*, 1 : 1-190.   *40*
[191] Jamieson, T. F. The 'parallel roads' of Glen Roy and their place in the history of the Glacial period. *QJGS*, 19 : 235-59. (See also 179, 183, 201, 216 and 361.)   *40*
[192] Lyell, C. *Geological evidences of the antiquity of Man.* London, Murray. L.   *39*
[193] Morton, G. H. *The geology of the country around Liverpool.* London, Philip. L.   *50*
[194] Ramsay, A. C. Breaks in succession of the British strata. *QJGS*, 19 : *(Proceedings)* xxxvi-lii; 20 : *(Proceedings)* xl-lx. 1863-64.   *52*
[195] Ramsay, A. C. *The physical geology and geography of Great Britain.* London, Stanford. L.   *53*

1864

[196] Prestwich, J. Theoretical considerations on the conditions under which the drift deposits containing the remains of extinct Mammalia and flint implements were accumulated . . . *Phil. Trans. Roy. Soc.*, 154 : 247-309   *39*
[197] Salter, J. W. *British trilobites.* London Palaeontogr. Soc., 1864-83.   *39*
[198] Wright, T. *The British fossil Echinodermata from the Cretaceous formations—Echinoidea.* London, Palaeontogr. Soc., 1864-82.   *39*

1865

[199] Foster, C. le N. and Topley, W. On the superficial deposits of the valley of the Medway with remarks on the denudation of the Weald. *QJGS*, 21 : 443-74.   *53*
[200] Geikie, A. *The scenery of Scotland.* London, Macmillan. L. *53*
[201] Jamieson, T. F. The history of the last glacial changes in Scotland. *QJGS*, 21 : 161-203. (See also 179, 183, 191, 216 and 361.)   *40*

1866

[202] Dawkins, W. B. *The British Pleistocene Mammalia.* London, Palaeontogr. Soc., 1866-72.   *39*
[203] Ramsay, A. C. and Salter, J. W. *The geology of North Wales.* London, Geol. Surv. L; 2nd, 1881.   *51*

1867

[204] Whitaker, W. On subaerial denudation, and on cliffs and escarpments of the Chalk and Lower Tertiary beds. *Geol. Mag.*, 4 : 447-54, 483-93.   *53*

1868

[205] Traquair, R. H., Powrie, J., and Lankester, E. R. *The fishes of the Old Red Sandstone of Britain.* London, Palaeontogr. Soc., 1868-1914.   *39*

1869

[206] Green, A. H. and others. *The geology of North Derbyshire.* London, Geol. Surv. L; 2nd, 1887.   *50*

1871

[207] Hicks, H. and Harkness, R. The ancient rocks of St. David's promontory. *QJGS*, 27 : 384-404. (See also 211.)   *51*

[208] Phillips, J. *Geology of Oxford and the valley of the Thames.* Oxford, Clarendon Press.   *50*

1872

[209] Evans, J. *Ancient stone-implements, weapons, and ornaments of Great Britain.* London, Longman. L.   *39*

[210] Whitaker, W. *The geology of the London basin.* London, Geol. Surv. (See also 280.)   *50*

1873

[211] Hicks, H. [The 'Pre-Cambrian' and Lower Palaeozoic rocks of Pembrokeshire, particularly those in the neighbourhood of St David's.] *QJGS*, 29 : 39-52; 31 : 167-95; 33 : 229-41; 34 : 153-69; 35 : 285-94, 40 : 507-60; 42 : 351-6. 1873-86. (See also 207.)   *51*

[212] Judd, J. W. The Secondary rocks of Scotland. Three papers, the second (1874) On the ancient volcanoes of the Highlands and the relations of their products to the Mesozoic strata. *QJGS*, 29 : 97-195; 30 : 220-302; 34 : 660-741. 1873-78. (See also 258, 263, 277, 281, 299, 300.)   *43, 71*

1874

[213] Allport, S. On the microscopic structure and composition of British Carboniferous dolerites. *QJGS*, 30 : 529-67.   *48*

[214] Dawkins, W. B. *Cave hunting.* London, Macmillan.   *39*

[215] Geikie, J. The Great Ice Age. London, Isbister. L.   *40*

[216] Jamieson, T. F. The last stage of the Glacial period in North Britain. *QJGS*, 30 : 317-37. (See also 179, 183, 191, 201 and 361.)   *40*

1875

[217] Blake, J. F. On the Kimmeridge Clay of England. *QJGS*, 31 : 196-237.   *71*

[218] Judd, J. W. *The geology of Rutland* [etc.] London, Geol. Surv.   *50*

[219] Topley, W. *The geology of the Weald*. London, Geol. Surv.   *50, 53*

[220] Ward, J. C. Notes on the comparative microscopic rock-structure of some ancient and modern volcanic rocks. *QJGS*, 31 : 388-422. (See also 221 and 226.)   *48, 73*

[221] Ward, J. C. The granitic, granitoid, and associated metamorphic rocks of the Lake-district. *QJGS*, 31 : 568-602; 32 : 1-34. 1875-6. (See also 220 and 226.)   *48, 73*

1876

[222] Allport, S. On the metamorphic rocks surrounding the Lands-End mass of granite. *QJGS*, 32 : 407-27.   *48*

[223] Blake, J. F. and Tate, R. *The Yorkshire Lias*. London, Van Voorst.   *71*

[224] Geikie, A. *Outlines of field geology*. London, Macmillan. L.   *38*

[225] Green, A. H. *Physical geology*. London, Rivington. L.   *38*

[226] Ward, J. C. *The geology of the northern part of the English Lake District*. London Geol. Surv. (See also 220 and 221.)   *73*

[227] Woodward, H. B. *The geology of England and Wales*. London, Philip. L; 2nd, 1887.   *38, 50*

1877

[228] Allport, S. On certain ancient devitrified pitchstones and perlites from the Lower Silurian district of Shropshire. *QJGS*, 33 : 449-60.   *48*

[229] Blake, J. F. and Hudleston, W. H. The Corallian rocks of England. *QJGS*, 33 : 260-405.   *71*

[230] Bonney, T. G. On the serpentine and associated rocks of the Lizard district. *QJGS*, 33 : 884-928. (See also 251 and 285.)   *50*

[231] Bonney, T. G. and Hill, E. The Precarboniferous rocks of Charnwood Forest. *QJGS*, 33 : 754-89; 34 : 199-239; 36 : 337-50. 1877-80. (See also 286.)   *50*

[232] Callaway, C. A new area of Upper Cambrian rock in South Shropshire. *QJGS*, 33 : 652-72. (See also 233, 238 and 287.)   50

1878

[233] Callaway, C. On the quartzites of Shropshire. *QJGS*, 34 : 754-63. (See also 232, 238 and 287.)   50
[234] Geikie, A. On the Old Red Sandstone of western Europe. *Trans. Roy. Soc. Edinb.*, 28 : 345-452.   45
[235] Green, A. H. *The geology of the Yorkshire coalfield*. London, Geol. Surv.   50
[236] Lapworth, C. The Moffat series. *QJGS*, 34 : 240-346.   57
[237] Newton, E. T. *The chimaeroid fishes from the British cretaceous rocks*. London, Geol. Surv.   39

1879

[238] Callaway, C. and Bonney, T. G. The Precambrian rocks of Shropshire. *QJGS*, 35 : 643-69, 38 : 119-26. 1797-82. (See also 232, 233 and 287.)   50
[239] Geikie, A. On the Carboniferous volcanic rocks of the basin of the Firth of Forth. *Trans. Roy. Soc. Edinb.*, 29 : 437-518.   43
[240] Lapworth, C. On the tripartite classification of the Lower Palaeozoic rocks [Proposal of the Ordovician system.] *Geol. Mag.*, 16 : 1-15.   36, 69
[241] Lapworth, C. On the geological distribution of the Rhabdophora rocks [graptolites]. *Ann. Mag. Nat. Hist.*, 5th series, 3 : 245-57, 449-55; 4 : 331-41; 5 : 45-62, 273-85, 358-69; 6 : 16-29, 185-207. 1879-80.   69
[242] Peach, B. N. and Horne, J. The glaciation of the Shetland Isles. *QJGS*, 35 : 778-811. (See also 246 and 248.)   40

1880

[243] Blake, J. F. On the Portland rocks of England. *QJGS*, 36 : 189-236.   71
[244] Bonney T. G. Petrological notes on the vicinity of the upper part of Loch Maree. *QJGS*, 36 : 93-108.   54
[245] Dawkins, W. B. *Early Man in Britain*. London, Macmillan.
                                                                                                                 39
[246] Peach, B. N. and Horne, J. The glaciation of the Orkney islands. *QJGS*, 36 : 648-63. (See also 242 and 248.)   40
[247] Wood, S. V., the younger. The newer Pliocene [including the Pleistocene] period in England. *QJGS*, 36 : 457-528; 38 : 667-745. 1880-2.   74

1881

[248] Peach B. N. and Horne, J. The glaciation of Caithness. *Proc. Roy. Phys. Soc. Edinb.*, 6 : 316-52. (See also 242 and 246.)
40

1882

[249] Geikie, A. *Text-book of geology*. London, Macmillan. L; 4th 1903. *38*

[250] Lapworth, C. The Girvan succession. *QJGS*, 38 : 537-666. *57*

1883

[251] Bonney, T. G. The hornblendic and other schists of the Lizard district. *QJGS*, 39 : 1-24. (See also 230 and 285.)
50

[252] Callaway, C. The age of the newer gneissic rocks of the northern Highlands. *QJGS*, 39 : 355-422. *54*

[253] Lapworth, C. The secret of the Highlands. *Geol. Mag.*, 20 : 120-8, 193-9, 337-44. (See also 259.) *54*

1884

[254] Bonney, T. G. On the Archaean rocks of Great Britain. *Rep. Br. Ass. Advmt. Sci.* for 1884, 529-51. *50*

[255] Hinde, G. J. *Catalogue of the fossil sponges of the British Museum*. London, British Museum. *39*

[256] Jukes-Browne, A. J. *The student's handbook of stratigraphical geology*. London, Stanford. L; 2nd, 1912. *38, 51*

1885

[257] Harker, A. On slaty cleavage and allied rock structures. *Rep. Br. Ass. Advmt. Sci.* for 1885, 813-52. *41*

[258] Judd, J. W. On the Tertiary and older peridotites of Scotland. *QJGS*, 41 : 354-418. (See also 212, 263, 277, 281, 299, 300.) *43*

[259] Lapworth, C. On the close of the Highland controversy. *Geol. Mag.*, 22 : 97-106. (See also 253.) *54*

[260] Marr, J. E. and Roberts, T. The Lower Palaeozoic rocks of the neighbourhood of Haverfordwest. *QJGS*, 41 : 476-91.
51

[261] Phillips, J., Seeley, H. G., and Etheridge, R. *Manual of geology*. London, Griffin. *38*

[262] Rutley, F. *The felsitic lavas of England and Wales*. London, Geol. Surv. *48*

1886
[263] Judd, J. W. On the gabbros, dolerites, and basalts, of Tertiary age, in Scotland and Ireland. *QJGS,* 42 : 49-97. (See also 212, 258, 277, 281, 299, 300.)   *43*
[264] Prestwich, J. *Geology—chemical, physical, and stratigraph- ical.* Oxford, Clarendon Press. 1886-8.   *38*

1887
[265] Hinde, G. J. *British fossil sponges.* London, Palaeontogr. Soc., 1887-1912.   *39*
[266] Hudleston, W. H. *Gasteropoda of the Inferior Oolite.* London, Palaeontogr. Soc., 1887-96.   *39*

1888
[267] Etheridge, R. *Fossils of the British Islands—Palaeozoic.* Oxford, Clarendon Press.   *39*
[268] Geikie, A. The history of volcanic action during the Tertiary period in the British Isles. *Trans. Roy. Soc. Edinb.,* 35 : 23-184.   *43*
[269] Geikie, A. Report on the recent work of the Geological Survey in the north-west Highlands of Scotland, based on the field-notes and maps of Messrs. B. N. Peach, J. Horne, W. Gunn, C. T. Clough, L. Hinxman, and H. M. Cadell. *QJGS,* 44 : 378-441.   *54*
[270] Jukes-Browne, A. J. *The building of the British Isles.* London, Stanford. L; 3rd, 1911; 4th, 1922.   *38, 67*
[271] Lapworth, C. On the discovery of the *Olenellus* fauna in the Lower Cambrian rocks of Britain. *Geol. Mag.,* 25 : 484-87. (See also 292.)   *50*
[272] Marr, J. E. and Nicholson, H. A. The Stockdale Shales. *QJGS,* 44 : 654-732. (See also 463.)   *73*
[273] Newton, E. T. On the skull, brain, and auditory organ of a new species of pterosaurian from the Upper Lias near Whitby, Yorkshire. *Phil. Trans. Roy. Soc.* (series B), 179 : 503-37.   *39*
[274] Teall, J. J. H. *British petrography.* London, Dulau.   *48*

1889
[275] Buckman, S. S. On the Cotteswold, Midford, and Yeovil Sands, and the division between the Lias and Oolite. *QJGS,* 45 : 440-73. (See also 298, 311, 316 and 339.)   *71*
[276] Harker, A. *The Bala volcanic series of Caernarvonshire.* Cambridge, University Press.   *51*
[277] Judd, J. W. The Tertiary volcanoes of the Western Isles of

Scotland. *QJGS*, 45 : 187-219. (See also 212, 258, 263, 281, 299 and 300.) *43*
[278] Lamplugh, G. W. On the subdivisions of the Speeton Clay. *QJGS*, 45 : 575-618. (See also 315.) *71*
[279] Whidborne, G. F. *The Devonian fauna of the south of England.* London, Palaeontogr. Soc., 1889-1907. *39*
[280] Whitaker, W. *The geology of London and of part of the Thames valley.* London, Geol. Surv. (See also 210.) *50*

1890
[281] Judd, J. W. The propylites of the Western Isles of Scotland, and their relation to the andesites and diorites of the district. *QJGS*, 46 : 341-85. (See also 212, 258, 263, 277, 299 and 300.) *43*
[282] Reid, C. *The Pliocene deposits of Britain.* London, Geol. Surv. *74*
[283] Ussher, W. A. E. The Devonian rocks of South Devon. *QJGS*, 46 : 487-517. *75*
[284] Woodward, A. S. and Sherborn, C. D. *A catalogue of British fossil Vertebrata.* London, Dulau. *39*

1891
[285] Bonney, T. G. Results of an examination of the crystalline rocks of the Lizard district. *QJGS*, 47 : 464-99. (See also 230 and 251.) *50*
[286] Bonney, T. G. and Hill, E. On the north-west region of Charnwood Forest. *QJGS*, 47 : 78-108. (See also 231.) *50*
[287] Callaway, C. The unconformities between the rock-systems underlying the Cambrian Quartzite of Shropshire. *QJGS*, 47 : 109-25. (See also 232, 233 and 238.) *50*
[288] Geikie, A. The history of volcanic action in the British Isles. (Anniversary addresses.) *QJGS*, 47 : *(Proceedings)* 63-162; 48 : *(Proceedings)* 60-179. 1891-2. *59*
[289] Harker, A. and Marr, J. E. The Shap Granite and the associated igneous and metamorphic rocks. *QJGS*, 47 : 266-328; 49 : 359-71. 1891-3. *73*
[290] Hatch, F. H. *Petrology of the igneous rocks.* London, Murby. L; rewritten 12th, with A. K. and M. K. Wells, 1961. *59, 61*
[291] Lamplugh, G. W. On the drifts of Flamborough Head. *QJGS*, 47 : 384-431. *40*
[292] Lapworth, C. On *Olenellus callavei* and its geological relationships. *Geol. Mag.*, 28 : 529-36. (See also 271.) *39, 50*

[293] Marr, J. E., Nicholson, H. A. and Harker, A. The Cross Fell inlier. *QJGS*, 37 : 500-29.   *70*

1892

[294] Fox-Strangways, C. E. *The Jurassic rocks of Britain—York-shire*. London, Geol. Surv.   *71*

[295] Peach, B. N. and Horne, J. The *Olenellus* zone in the north-west Highlands of Scotland. *QJGS*, 48 : 227-42.   *54*

[296] Prestwich, J. The raised beaches of the south of England. *QJGS*, 48 : 263-343.   *74*

1893

[297] Barrow, G. An intrusion of muscovite-biotite gneiss in the south-eastern Highlands of Scotland, and its accompanying metamorphism. *QJGS*, 49 : 330-58.   *55*

[298] Buckman, S. S. The Bajocian of the Sherborne district. *QJGS*, 49 : 479-522. (See also 275, 311, 316 and 339.)   *71*

[299] Judd, J. W. On inclusions of Tertiary granite in the gabbro of the Cuillin Hills, Skye. *QJGS*, 49 : 175-95. (See also 212, 258, 263, 277, 281, 300.)   *43*

[300] Judd, J. W. On composite dykes in Arran. *QJGS*, 49 : 536-65. (See also 212, 258, 263, 277, 281, 299.)   *43*

[301] Newton, E. T. Reptiles from the Elgin Sandstone. *Phil. Trans. Roy Soc.* (series B) 184 : 431-503; 185 : 573-607. 1893-4.
                                                                        39

[302] Woods, H. *Palaeontology: invertebrate*. Cambridge, University Press. L; 8th, 1946.   *63*

[303] Woodward, H. B. *The Jurassic rocks of Britain—Yorkshire excepted*. London, Geol. Surv., 1893-5.   *71*

1894

[304] Harker, A. Carrock Fell : a study in the variation of igneous rock-masses. *QJGS*, 50 : 311-37; 51 : 125-48. 1894-5.   *73*

[305] Hind, W. *Carbonicola, Anthracomya,* and *Naiadites*. London, Palaeontogr. Soc., 1894-6. (See also 314.)   *39*

[306] Kidston, R. On the various divisions of the British Carbon-iferous rocks as determined by their fossil flora. *Proc. Roy. Phys. Soc. Edinb.*, 12 : 183-257.   *39*

[307] Lapworth, C. and Watts, W. W. The geology of South Shropshire. *Proc. Geol. Ass.*, 13 : 297-355. L; in *Geology in the field*, 1910.   *50*

[308] Lewis, H. C. *The glacial geology of Great Britain and Ireland*. Edited by H. W. Crosskey. London, Macmillan.   *40*

[309] Newton, E. T. The [Pleistocene] vertebrate fauna from the fissure near Ightham, Kent. *QJGS*, 54 : 188-211.   *39*

[310] Seward, A. C. *The Wealden flora*. London, British Museum
(Natural History), 1894-5. *39*

1895

[311] Buckman, S. S. The Bajocian of the Mid-Cotteswolds. *QJGS,*
51 : 388-462. (See also 275, 298, 316 and 339.) *71*
[312] Harker, A. *Petrology for students*. Cambridge, University
Press. L; 8th, 1954. *61*

1896

[313] Geikie, A. The Tertiary basalt-plateaux of north-western
Europe. *QJGS*, 52 : 331-406. *43*
[314] Hind, W. *British Carboniferous Lamellibranchiata*. London,
Palaeontogr. Soc., 1896-1905. (See also 305.) *39*
[315] Lamplugh, G. W. On the Speeton series in Yorkshire and
Lincolnshire. *QJGS*, 52 : 179-220. (See also 278.) *71*

1897

[316] Buckman, S. S. Deposits of the Bajocian age in the northern
Cotteswolds : the Cleeve Hill plateau. *QJGS*, 53 : 607-29.
(See also 275, 298, 311 and 339.) *71*
[317] Clough, C. T. *The geology of Cowal*. London, Geol. Surv.
54
[318] Geikie, A. *The ancient volcanoes of Great Britain*. London,
Macmillan. *23, 43, 59*
[319] Geikie, A. *Geological map of England and Wales*. (Scale, 10
miles to 1 inch.) London. *49*

1898

[320] Elles, G. L. The graptolite fauna of the Skiddaw Slates. *QJGS,*
54 : 463-539. *69, 73*
[321] Geikie, J. *Earth sculpture*. London, Murray. *53*
[322] Harmer, F. W. The Pliocene deposits of the east of England.
QJGS, 54 : 308-56; 56 : 705-44. 1898-1900. *74*
[323] Lapworth, C., Watts, W. W., and Harrison, W. J. A sketch
of the geology of the Birmingham district. *Proc. Geol. Ass.,*
15 : 313-416. *50*
[324] Seward, A. C. *Fossil plants*. Cambridge, University Press.
1898-1919. *39*
[325] Strahan, A. *The geology of the Isle of Purbeck and
Weymouth*. London, Geol. Surv. *26*

1899

[326] Groom, T. T. The geological structure of the southern Malverns, and the adjacent district to the west. *QJGS*, 55 : 129-69. (See also 334, 342 and 347.)   *46*

[327] Peach, B. N. and Horne, J. *The Silurian rocks of Scotland.* London, Geol. Surv.   *57*

[328] Rowe, A. W. An analysis of the genus *Micraster*, as determined by rigid zonal collecting. *QJGS*, 55 : 494-547.   *64*

[329] Strahan, A. and others. *The geology of the South Wales coalfield.* London, Geol. Surv., 1899-1921.   *77*

[330] Traquair, R. H. Fossil fishes in the Silurian rocks of South Scotland. *Trans. Roy. Soc. Edinb.*, 39 : 591-4. 1899-1905.   *39*

[331] Whitaker, W. *The water supply of Sussex* (1899), *Kent* (1908), and *Buckinghamshire* and *Hertfordshire* (1921). London, Geol. Surv., 1899-1921.   *42*

[332] Woods, H. *Cretaceous Lamellibranchia of England.* London, Palaeontogr. Soc., 1899-1913. (See also 396.)   *39*

1900

[333] Elles, G. L. The zonal classification of the Wenlock Shales of the Welsh Borderland. *QJGS*, 56 : 370-414.   *69*

[334] Groom, T. T. On the geological structure of portions of the Malvern and Abberley Hills. *QJGS*, 56 : 138-97. (See also 326, 342 and 347.)   *46*

[335] Jukes-Browne, A. J. and Hill, W. *The Cretaceous rocks of Britain.* London, Geol. Surv., 1900-4.   *71*

[336] Rowe, A. W. The zones of the White Chalk of the English coast. *Proc. Geol. Ass.*, 16 : 289-367; 17 : 1-76; 18 : 1-51; 19 : 193-296; 20 : 209-328, 336-9. 1900-8.   *71*

[337] Seward, A. C. *The Jurassic flora.* London, British Museum (Natural History), 1900-4.   *62*

[338] Wood, E. M. R. The Lower Ludlow formation and its graptolite fauna. *QJGS*, 56 : 415-92.   *69*

1901

[339] Buckman, S. S. The Bajocian and contiguous deposits of the North Cotteswolds. *QJGS*, 57 : 126-55. (See also 275, 298, 311, and 316.)   *71*

[340] Elles, G. L. and Wood, E. M. R. *British graptolites.* London, Palaeontogr. Soc., 1901-19.   *69*

[341] Gibson, W. On the character of the Upper Coal-Measures of North Staffordshire, Denbighshire, South Staffordshire, and Nottinghamshire. *QJGS*, 57 : 251-66.   *77*

[342] Groom, T. T. On the igneous rocks associated with the Cambrian beds of the Malvern Hills. *QJGS*, 57 : 156-84. (See also 326, 334 and 347.) *46*

[343] Harker, A. Ice-erosion in the Cuillin Hills, Skye. *Trans. Roy. Soc. Edinb.*, 40 : 221-52. *74*

[344] Hind, W. and Howe, J. A. The geological succession and palaeontology of the beds between the Millstone Grit and the Limestone-massif at Pendle Hill and their equivalents in certain other parts of Britain. *QJGS*, 57 : 347-404. *77*

[345] Hughes, T. McK. Ingleborough. *Proc. Yorks. Geol. Soc.*, 14 : 125-50, 323-43; 15 : 351-71; 16 : 45-74, 177-96, 253-320. 1901-8. *70*

1902

[346] Geikie, A. *The geology of East Fife*. London, Geol. Surv. *56*

[347] Groom, T. T. The sequence of the Cambrian and associated beds of the Malvern Hills. *QJGS*, 58 : 89-149. (See also 326, 334 and 342.) *46*

[348] Kendall, P. F. A system of glacier lakes in the Cleveland Hills. *QJGS*, 58 : 471-569. *74*

[349] Woodward, A. S. *Fishes of the English Chalk*. London, Palaeontogr. Soc., 1902-12. *62*

1903

[350] Lamplugh, G. W. *The geology of the Isle of Man*. London, Geol. Surv. *24, 72*

[351] Reed, F. R. C. *The Lower Palaeozoic trilobites of the Girvan district, Ayrshire*. London, Palaeontogr. Soc., 1903-35. *62*

[352] Traquair, R. H. Distribution of fossil fish remains in the Carboniferous of the Edinburgh district. *Trans. Roy. Soc. Edinb.*, 40 : 687-707. *62*

[353] Ussher, W. A. E. *The geology of the country around Torquay*. (And several other districts of SW. England.) London, Geol. Surv. *75*

1904

[354] Arber, E. A. N. The fossil flora of the Culm Measures of north-west Devon. *Phil. Trans. Roy. Soc.* (series B), 197 : 291-325. (See also 391 and 411.) *62*

[355] Harker, A. *The Tertiary igneous rocks of Skye*. London, Geol. Surv. *58*

[356] Richardson, L. *A handbook to the geology of Cheltenham and neighbourhood*. Cheltenham, Norman and Sawyer. *71*

1905

[357] Fearnsides, W. G. On the geology of Arenig Fawr and Moel Llyfnant. *QJGS*, 61 : 608-40. (See also 380 and 573.)   *78*

[358] Geikie, J. *Structural and field geology*. Edinburgh, Oliver and Boyd. L; 6th, with R. Campbell and R. M. Graig, 1952.   *65*

[359] Gibson, W. *The geology of the North Staffordshire coalfields*. London, Geol. Surv.   *77*

[360] Vaughan, A. The palaeontological succession in the Carboniferous Limestone of the Bristol area. *QJGS*, 61 : 181-307.   *76*

1906

[361] Jamieson, T. F. The glacial period in Aberdeenshire and the southern border of the Moray Firth. *QJGS*, 62 : 13-39. (See also 179, 183, 191, 201 and 216.)   *40*

[362] Lake, P. *British Cambrian trilobites*. London, Palaeontogr. Soc., 1906-46.   *62*

[363] Marr, J. E. The influence of the geological structure of English Lakeland upon its present features. *QJGS*, 62 : *(Proceedings)* lxvi-cxxviii.   *73*

[364] Teall, J. J. H. (director) Geological map of the British Islands. (Scale, 25 miles to 1 inch.) London, Geol. Surv. L.   *68*

1907

[365] Bemrose, H. H. A. The toadstones of Derbyshire. *QJGS*, 63 : 241-81.   *60*

[366] Peach, B. N., Horne, J., and others. *The geological structure of the north-west Highlands of Scotland*. London, Geol. Surv.   *54*

[367] Richardson, L. The Inferior Oolite and contiguous deposits of the Bath-Doulting district and of the district between the Rissingtons and Burford. *QJGS*, 63 : 383-426, 437-44. (See also 417.)   *71*

1908

[368] Green, J. F. N. The geological structure of the St. David's area (Pembrokeshire). *QJGS*, 64 : 363-83.   *51*

[369] Harker, A. *The geology of the Small Isles of Inverness-shire*. London, Geol. Surv.   *58*

[370] Sibly, T. F. The faunal succession in the Carboniferous Limestone (Avonian) of the Midland area (North Derbyshire and North Staffordshire). *QJGS*, 64 : 34-82.   *76*

1909

[371] Buckman, S. S. *Type ammonites* and *Yorkshire type ammonites.* London, Murby. 1909-30.   62

[372] Clough, C. T., Maufe, H. B., and Bailey, E. B. The cauldron-subsidence of Glen Coe, and the associated igneous phenomena. *QJGS*, 65 : 611-78.   55

[373] Davis, W. M. Glacial erosion in North Wales. *QJGS*, 65 : 281-350.   74

[374] Harker, A. *The natural history of igneous rocks.* London, Methuen.   61

[375] Jones, O. T. The Hartfell-Valention succession in the district around Plynlimon and Pont Erwyd (North Cardiganshire). *QJGS*, 65 : 463-537.   79

[376] Monckton, H. W. and Herries, R. S. (editors). *Geology in the field* [England and Wales]: *The Jubilee Volume of the Geologists' Association.* 1909-10.   *46, 50, 66, 78*

1910

[377] Bailey, E. B. Recumbent folds in the schists of the Scottish Highlands. *QJGS*, 66 : 586-620.   55

[378] Carruthers, R. G. On the evolution of *Zaphrentis delanouei* in Lower Carboniferous times. *QJGS*, 66 : 523-38.   64

[379] Clough, C. T. and others. *The geology of East Lothian.* London, Geol. Surv.   56

[380] Fearnsides, W. G. The Tremadoc Slates and associated rocks of south-east Caernarvonshire. *QJGS*, 66 : 142-88. (See also 357 and 573.)   78

[381] Flett, J. S. Petrological chapter in *The geology of the neighbourhood of Edinburgh.* 2nd edition. (And petrological chapters in several other Scottish memoirs.) London, Geol. Surv.   60

[382] Lake, P. and Rastall, R. H. *A textbook of geology.* London, Arnold. L; 5th, 1941.   65

[383] Rastall, R. H. The Skiddaw granite and its metamorphism. *QJGS*, 66 : 116-41.   73

1911

[384] Arber, E. A. N. *The coast scenery of North Devon.* London, Dent.   *53, 72*

[385] Dewey, H. and Flett, J. S. British pillow-lavas and the rocks associated with them. *Geol. Mag.*, 48 : 202-9, 241-8.   60

[386] Dixon, E. E. L. and Vaughan, A. The Carboniferous succession in Gower (Glamorganshire), with notes on its fauna and conditions of deposition. *QJGS*, 67 : 477-571.   76

[387] Holmes, A. The association of lead with uranium in rock-minerals, and its application to the measurement of geological time. *Proc. Roy. Soc.* (series A), 85 : 248-56. (See also 588 and 639.)   *81*

[388] Sherlock, R. L. The relationship of the Permian to the Trias in Nottinghamshire. *QJGS*, 67 : 75-117.   *80*

[389] Sollas, W. J. *Ancient hunters.* London, Macmillan. L.   *39*

[390] Thomas, H. H. The Skomer volcanic series (Pembrokeshire). *QJGS*, 67 : 175-214.   *60*

1912

[391] Arber, E. A. N. The fossil flora of the Forest of Dean coalfield. *Phil. Trans. Roy. Soc.* (series B), 202 : 233-81. (See also 354 and 411.)   *62*

[392] Flett, J. S. *The geology of the Lizard and Meneage.* London, Geol. Surv. L.   *72*

[393] Garwood, E. J. The Lower Carboniferous succession in the north-west of England. *QJGS*, 68 : 449-586.   *76*

[394] Jones, O. T. The geological structure of Central Wales and the adjoining regions. *QJGS*, 68 : 326-44.   *51, 79*

[395] Smith, B. The glaciation of the Black Combe district (Cumberland). *QJGS*, 68 : 402-8.   *74*

[396] Woods, H. The evolution of *Inoceramus* in the Cretaceous period. *QJGS*, 68 : 1-20. (See also 332.)   *64*

1913

[397] Bailey, E. B. The Loch Awe syncline (Argyllshire). *QJGS*, 69 : 280-307.   *55*

[398] Davies, A. M. and Pringle, J. Two deep borings at Calvert station (North Buckinghamshire) and the Palaeozoic floor north of the Thames. *QJGS*, 69 : 308-42.   *82*

[399] Hatch, F. H. and Rastall, R. H. *Petrology of the sedimentary rocks.* London, Murby. L.   *61*

[400] Matley, C. A. The geology of Bardsey Island. *QJGS*, 69 : 514-33. (See also 488, 503, 517, 551, and 563.)   *78*

[401] Reid, C. *Submerged forests.* Cambridge, University Press. *74*

[402] Trechmaun, C. T. A mass of anhydrite in the Magnesian Limestone at Hartlepool, and on the Permian of south-eastern Durham. *QJGS*, 69 : 184-218. (See also 406 and 466.)   *80*

1914

[403] Crampton, C. B. and Carruthers, R. G. *The geology of Caithness.* London, Geol. Surv.   *75*

[404] Harmer, F. W. *The Pliocene Mollusca of Great Britain.*
London, Palaeontogr. Soc., 1914-24. *62*

[405] Lang, W. D. The Lower Lias of the Dorset coast (and adjacent parts). *Proc. Geol. Ass.,* 25 : 293-360; 28 : 30-6;
35 : 169-85; 43 : 97-126. 1914-32. *71*

[406] Trechmann, C. T. The lithology and composition of Durham
Magnesian limestones. *QJGS,* 70 : 232-65. (See also 402
and 466.) *80*

[407] Wright, W. B. *The Quaternary ice age.* London, Macmillan.
L. *74*

1915

[408] Nicholas, T. C. The geology of the St. Tudwal's peninsula
(Caernarvonshire). *QJGS,* 71 : 83-143, 451-72. 1915-16.
72, 78

[409] Trenchmann, C. T. The Scandinavian Drift of the Durham
coast. *QJGS,* 71 : 53-82. *74*

[410] Vaughan, A. Correlation of Dinantian and Avonian. *QJGS,*
71 : 1-52. *76*

1916

[411] Arber, E. A. N. The fossil floras of the Coal Measures of
South Staffordshire. *Phil. Trans. Roy. Soc.* (series B), 208 :
127-55. (See also 354 and 391.) *62*

[412] Bailey, E. B. *The geology of Ben Nevis and Glen Coe and the
surrounding country.* London, Geol. Surv. L; 2nd, 1960.
55

[413] Boswell, P. G. H. The stratigraphy and petrology of the
Lower Eocene deposits of the north-eastern part of the
London Basin. *QJGS,* 71 : 536-91. *72*

[414] Illing, V. C. The paradoxidian fauna of a part of the Stockingford Shales. *QJGS,* 71 : 386-450. *72*

[415] Jones, O. T. and Pugh, W. J. The geology of the district
around Machynlleth and the Llyfnant valley. *QJGS,* 71 :
343-85. *79*

[416] Marr, J. E. *The geology of the Lake District.* Cambridge,
University Press. *30, 73*

[417] Richardson, L. The Inferior Oolite and contiguous deposits
of the Doulting-Milborne-Port district and of the Crewkerne district. *QJGS,* 71 : 473-520; 74 : 145-73. 1916-18.
(See also 367.) *71*

[418] Woodward, A. S. *The fossil fishes of the English Wealden
and Purbeck formations.* London, Palaeontogr. Soc., 1916-
19. *62*

1917
[419] Kidston, R. and Lang, W. H. On Old Red Sandstone plants
showing structure. *Trans. Roy. Soc. Edinb.*, 51 : 761-84;
52 : 603-27, 643-50, 831-902. 1917-21.   62
[420] Reed, F. R. C. The Ordovician and Silurian Brachiopoda of
the Girvan district. *Trans. Roy. Soc. Edinb.*, 51 : 795-998.
                                                           62
[421] Tyrrell, G. W. The picrite-teschenite sill of Lugar (Ayrshire). *QJGS*, 72 : 84-131.   60

1918
[422] Harker, A. Some aspects of igneous activity in Britain. *QJGS*,
73 : *(Proceedings)* lxvii-xcvi.   59
[423] Sibly, T. F. and Dixey, F. The Carboniferous Limestone series
on the south-eastern margin of the South Wales coalfield.
QJGS, 73 : 111-64.   76

1919
[424] Greenly, E. *The geology of Anglesey.* London, Geol. Surv.
                                                         34, 78

1920
[425] Davies, A. M. *Introduction to palaeontology.* London, Murby.
L; 3rd, with C. J. Stubblefield, 1961.   63
[426] Gilligan, A. The petrography of the Millstone Grit of Yorkshire. *QJGS*, 75 : 251-94.   77
[427] Lamplugh, G. W. Some features of the Pleistocene glaciation
of England. *QJGS*, 76 : *(Proceedings)* lxi-lxxxiii.   74

1921
[428] Bolton, H. *The fossil insects of the British Coal Measures.*
London, Palaeontogr. Soc., 1921-2.   62
[429] Cox, A. H. and Wells, A. K. The Lower Palaeozoic rocks of
the Arthog-Dolgelley district. *QJGS*, 76 : 254-324. (See
also 460.)   78
[430] Holmes, A. *Petrographic methods and calculations.* London,
Murby.   61
[431] Jones, O. T. The Valentian series. *QJGS*, 77 : 144-74. 79
[432] Sherlock, R. L. *Rock-salt and brine.* London, Geol. Surv.
(Mineral Resources.)   83
[433] Stamp, L. D. On cycles of sedimentation in the Eocene strata
of the Anglo-Franco-Belgian basin. *Geol. Mag.*, 63 : 108-
14, 146-57, 194-200.   72

1922

[434] Bailey, E. B. The structure of the south-west Highlands of Scotland. *QJGS*, 78 : 82-127.   55
[435] Elles, G. L. The graptolite faunas of the British Isles. *Proc. Geol. Ass.*, 33 : 168-200.   69
[436] Kidston, R. *Fossil plants of the Carboniferous rocks of Great Britain.* London, Geol. Surv. *(Palaeontology.)* 1922-6.   62
[437] Milner, H. B. *Sedimentary petrography.* London, Murby. L; 4th, 1962.   61
[438] Wills, L. J. and Smith, B. The Lower Palaeozoic rocks of the Llangollen district. *QJGS*, 78 : 176-226.   78
[439] Trueman, A. E. The use of *Gryphaea* in the correlation of the Lower Lias. *Geol. Mag.*, 59 : 256-68.   64

1923

[440] Anderson, E. M. The geology of the schists of the Schiehallion district. *QJGS*, 79 : 423-45.   55
[441] Boswell, P. G. H. The petrography of the Cretaceous and Tertiary outliers of the west of England. *QJGS*, 79 : 205-30.   72
[442] Greenly, E. Further researches on the succession and metamorphism in the Mona Complex of Anglesey. *QJGS*, 79 : 334-51. (See also 500.)   78
[443] Jehu, T. J. and Craig, R. M. The geology of the Outer Hebrides. *Trans. Roy. Soc. Edinb.*, 53 : 419-41, 615-41; 54 : 46-89; 55 : 457-88; 57 : 839-74. 1923-34.   54
[444] Lamplugh, G. W., Kitchin, F. L., and Pringle, J. *The concealed Mesozoic rocks of Kent.* London, Geol. Surv.   82
[445] Lang, W. D., Spath, L. F. and Richardson, W. A. Shales-with-'beef', a sequence in the Lower Lias of the Dorset coast. *QJGS*, 79 : 47-99. (See also 471 and 487.)   71
[446] Pugh, W. J. The geology of the district around Corris and Aberllefenni. *QJGS*, 79 : 508-41. (See also 490 and 495.)   78
[447] Read, H. H. *The geology of the country around Banff, Huntly and Turriff.* London, Geol. Surv.   55
[448] Read, H. H. The petrology of the Arnage district, Aberdeenshire. *QJGS*, 79 : 446-86. (See also 481, 542, 614, 615 and 631.)   55
[449] Spath, L. F. *The Ammonoidea of the Gault.* London, Palaeontogr. Soc., 1923-43.   62
[450] Swinnerton, H. H. *Outlines of palaeontology.* London, Arnold. L; 3rd, 1947.   63

1924
[451] Bailey, E. B., Thomas, H. H., and others. *The Tertiary and Post-Tertiary geology of Mull, Loch Aline and Oban.* London, Geol. Surv. (See also 459.)   *58*

[452] Bisat, W. S. The Carboniferous goniatites of the north of England and their zones. *Proc. Yorks. Geol. Soc.*, 20 : 40-124.   *77*

[453] Boulton, W. S. A recently-discovered breccia-bed underlying Nechells (Birmingham) and its relations to the red rocks of the district. *QJGS*, 80 : 343-73. (See also 521.)   *80*

[454] Davison, C. *A history of British earthquakes.* Cambridge, University Press.   *83*

[455] Hudson, R. G. S. On the rhythmic succession of the Yoredale series in Wensleydale. *Proc. Yorks. Geol. Soc.*, 20 : 125-35.   *76*

[456] Kendall, P. F. and Wroot, H. E. *The geology of Yorkshire.* Leeds, Kendall and Wroot.   *72*

[457] Tilley, C. E. Contact metamorphism in the Comrie area of the Perthshire Highlands. *QJGS*, 80 : 22-71.   *55*

[458] Wills, L. J. The development of the Severn valley in the neighbourhood of Ironbridge and Bridgnorth. *QJGS*, 80 : 274-314.   *74*

1925
[459] Bailey, E. B. and Lee, G. W. *The Pre-Tertiary geology of Mull, Loch Aline and Oban.* London, Geol. Surv. (See also 451.)   *55*

[460] Cox, A. H. The geology of the Cader Idris range. *QJGS*, 81 : 539-94. (See also 429.)   *78*

[461] Jones, O. T. and Andrew, G. The geology of the Llandovery district. The southern area; the Llandovery rocks of Garth (Breconshire); and the relations between the Llandovery rocks of Llandovery and Garth. *QJGS*, 81 : 344-88, 389-406, 407-16. (See also 602.)   *79*

[462] Macgregor, M. Contribution to *The geology of the Glasgow district* 3rd edition. (And contributions to other Scottish *Memoirs.*) London, Geol. Surv.   *56*

[463] Marr, J. E. Conditions of deposition of the Stockdale Shales of the Lake District. *QJGS*, 81 : 113-36. (See also 272.)   *44*

[464] Read, H. H. *The geology of the country around Golspie, Sutherlandshire.* London, Geol. Surv.   *55*

[465] Tilley, C. E. A preliminary survey of metamorphic zones in the southern Highlands of Scotland. *QJGS*, 81 : 100-10.   *55*

[446] Trechmaun, C. T. The Permian formation in Durham. *Proc. Geol. Ass.*, 36 : 135-45. (See also 402 and 406.)   *80*
[467] Watts, W. W. and others. The geology of South Shropshire. *Proc. Geol. Ass.*, 36 : 321-405.   *84*

1926

[468] Davies, K. A. The geology of the country between Drygarn and Abergwesyn (Breconshire). *QJGS*, 82 : 436-64. (See also 486 and 523.)   *79*
[469] Fearnsides, W. G. and Morris, T. O. The stratigraphy and structure of the Cambrian slate-belt of Nantlle, Caernarvonshire. *QJGS*, 82 : 250-303.   *78*
[470] Gibson, W. *The concealed coalfield of Yorkshire and Nottinghamshire*. London, Geol. Surv. L.   *47*
[471] Lang, W. D. and Spath, L. F. The Black Marl of Black Ven and Stonebarrow, in the Lias of the Dorset coast. *QJGS*, 82 : 144-87. (See also 445 and 487.)   *71*
[472] Reynolds, S. H. Progress in the study of the Lower Carboniferous (Avonian) rocks of England and Wales. *Rep. Br. Ass. Advmt. Sci.* for 1926, 65-101.   *76*
[473] Sherlock, R. L. A correlation of the British Permo-Triassic rocks. *Proc. Geol. Ass.*, 37 : 1-72; 39 : 49-95. 1926-8.   *80*
[474] Tyrrell, G. W. *The Principles of petrology*. London. L; 2nd, 1929.   *61*

1927

[475] Boswell, P. G. H. *The geology of the country around Ipswich* (1927), around *Woodbridge*, around *Felixstowe and Orford* (1928), and around *Sudbury* (1929). London, Geol. Surv., 1927-9.   *74*
[476] Carruthers, R. G. and others. *The oil-shales of the Lothians.* London, Geol. Surv.   *56*
[477] Cobbold, E. S. The stratigraphy and geological structure of the Cambrian area of Comley (Shropshire). *QJGS*, 83 : 551-73.   *84*
[478] George, T. N. The Carboniferous Limestone (Avonian) succession of a portion of the north crop of the South Wales coalfield. *QJGS*, 83 : 38-95. (See also 525 and 568.)   *76*
[479] Heard, A. Old Red Sandstone plants showing structure from Brecon. *QJGS*, 83 : 195-209. (See also 566.)   *62*
[480] Hickling, H. G. A. *Sections of strata of the Coal Measures of Lancashire*. Newcastle-upon-Tyne, Lancashire and Cheshire Coal Association.   *77*

[481] Read, H. H. The igneous and metamorphic history of Cromar, Deeside, Aberdeenshire. *Trans. Roy. Soc. Edinb.*, 55 : 317-53. (See also 448, 542, 614, 615 and 631.)   *55*

[482] Stubblefield, C. J. and Bulman, O. M. B. The Shineton Shales of the Wrekin district. *QJGS*, 83 : 96-146.   *84*

[483] Trueman, A. E. and Davies, J. H. A revision of non-marine Lamellibranchiata of the Coal Measures. *QJGS*, 83 : 210-59.   *77*

[484] Williams, H. The geology of Snowdon. *QJGS*, 83 : 346-431. (See also 512.)   *78*

1928

[485] Allan, D. A. The geology of the Highland Border from Tayside to Noranside. *Trans. Roy. Soc. Edinb.*, 56 : 57-88. (See also 567.)   *75*

[486] Davies, K. A. Contributions to the geology of Central Wales. *Proc. Geol. Ass.*, 39 : 157-68. (See also 468 and 523.)   *79*

[487] Lang, W. D., Spath, L. F., and others. The Belemnite Marls of Charmouth, a series in the Lias of the Dorset coast. *QJGS*, 84 : 179-257. (See also 445 and 471.)   *71*

[488] Matley, C. A. The Pre-Cambrian complex and associated rocks of south-western Lleyn (Caernarvonshire). *QJGS*, 84 : 440-504. (See also 400, 503, 517, 551 and 563.)   *78*

[489] Neaverson, E. *Stratigraphical palaeontology*. Oxford, Clarendon Press. L; 2nd, 1955.   *63*

[490] Pugh, W. J. The geology of the district around Dinas Mawddwy. *QJGS*, 84 : 345-81. (See also 446 and 495.)   *78*

[491] Tyrrell, G. W. *The geology of Arran*. London, Geol. Surv.   *58*

[492] Whittard, W. F. The stratigraphy of the Valentian rocks of Shropshire. *QJGS*, 83 : 737-58; 88 : 859-902. 1928-32.   *84*

1929

[493] Arkell, W. J. *British Corallian Lamellibranchia*. London, Palaeontogr. Soc., 1929-37.   *62, 71*

[494] Evans, J. W. and Stubblefield, C. J. (editors). *Handbook of the geology of Great Britain*. London.   *50, 59, 66*

[495] Pugh, W. J. The geology of the district between Llanymawddwy and Llanuwchllyn. *QJGS*, 85 : 242-306. (See also 446 and 490.)   *78*

[496] Trotter, F. M. The glaciation of eastern Edenside, the Alston block, and the Carlisle plain. *QJGS*, 85 : 549-612.   *74*

[497] Wills, L. J. *The physiographical evolution of Britain*. London, Arnold.   67

1930

[498] Bailey, E. B. New light on sedimentation and tectonics. *Geol. Mag.*, 67 : 77-92.   83
[499] Bolton, H. Fossil insects from the South Wales coalfield. *QJGS*, 86 : 9-49.   62
[500] Greenly, E. Foliation and its relation to folding in the Mona Complex at Rhoscolyn. *QJGS*, 86 : 169-90. (See also 442.)   78
[501] Lewis, H. P. The Avonian succession in the south of the Isle of Man. *QJGS*, 86 : 234-90.   76
[502] Macgregor, M. Scottish Carboniferous stratigraphy. *Trans. Geol. Soc. Glasg.*, 18 : 442-558.   56
[503] Matley, C. A. and Heard, A. The geology of the country around Bodfean (south-western Caernarvonshire). *QJGS*, 86 : 130-68. (See also 400, 488, 517, 551 and 563.)   78
[504] Peach, B. N. and Horne, J. *Chapters on the geology of Scotland*. Oxford, University Press.   54
[505] Richey, J. E., Thomas, H. H., and others. *The geology of Ardnamurchan, North-west Mull, and Coll*. London, Geol. Surv.   58

1931

[506] Dix, E. The flora of the upper portion of the Coal Measures of North Staffordshire. *QJGS*, 87 : 160-79. (See also 507.)   77
[507] Dix, E. and Trueman, A. E. Some non-marine lamellibranchs of the upper part of the Coal Measures. *QJGS*, 87 : 180-211. (See also 506.)   77
[508] Garwood, E. J. The Tuedian beds of northern Cumberland and Roxburghshire east of the Liddel Water. *QJGS*, 87 : 97-159.   76
[509] Hollingworth, S. E. The glaciation of western Edenside and adjoining areas and the drumlins of Edenside and the Solway basin. *QJGS*, 87 : 281-359.   74
[510] Read, H. H. *The geology of central Sutherland*. London, Geol. Surv.   55
[511] Seward, A. C. *Plant life through the ages*. Cambridge, University Press.   63
[512] Williams, H. and Bulman, O. M. B. The geology of the Dolwyddelan syncline. *QJGS*, 87 : 425-58. (See also 484.)   78

1932

[513] Carruthers, R. G. and others. *The geology of the Cheviot Hills.* London, Geol. Surv.   72
[514] Dewey, H. The Palaeolithic deposits of the lower Thames valley. *QJGS*, 88 : 35-56.   74
[515] Harker, A. *Metamorphism.* London, Methuen. L; 3rd, 1950.
65
[516] Lee, G. W. and Pringle, J. A synopsis of the Mesozoic rocks of Scotland. *Trans. Geol. Soc. Glasg.*, 19 : 158-224.   71
[517] Matley, C. A. The geology of the country around Mynydd Rhiw and Sarn, south-western Lleyn, Caernarvonshire. *QJGS*, 88 : 238-73. (See also 400, 488, 503, 551 and 563.)
78
[518] Richey, J. E. Tertiary ring structures in Britain. *Trans. Geol. Soc. Glasg.*, 19 : 42-140.   58

1933

[519] Arkell, W. J. *The Jurassic system in Great Britain.* Oxford, Clarendon Press.   71
[520] Boswell, P. G. H. *On the mineralogy of sedimentary rocks.* London, Murby.   61
[521] Boulton, W. S. The rocks between the Carboniferous and Trias in the Birmingham district. *QJGS*, 89 : 53-82. (See also 453.)   80
[522] Cobbold, E. S. and Pocock, R. W. The Cambrian area of Rushton (Shropshire). *Phil. Trans. Roy. Soc.* (series B), 223 : 305-409.   84
[523] Davies, K. A. The geology of the country between Abergwesyn (Breconshire) and Pumpsaint (Carmarthenshire). *QJGS*, 89 : 172-201. (See also 468 and 486.)   79
[524] Fearnsides, W. G. A correlation of structures in the coalfields of the Midland Province. *Rep. Br. Ass. Advmt. Sci.* for 1933, 57-80.   77
[525] George, T. N. The Carboniferous Limestone series in the west of the Vale of Glamorgan. *QJGS*, 89 : 221-72. (See also 478 and 568.)   76
[526] Trueman, A. E. A suggested correlation of the Coal Measures of England and Wales. *Proc. S. Wales Inst. Engrs.*, 49 : 63-106.   77

1934

[527] Bailey, E. B. West Highland tectonics : Loch Leven to Glen Roy. *QJGS*, 90 : 462-523.   55
[528] Davies, A. M. *Tertiary faunas.* London, Murby. 1934-5.   63

[529] Dix, E. The sequence of floras in the Upper Carboniferous, with special reference to South Wales. *Trans. Roy. Soc. Edinb.*, 57 : 789-838.   77
[530] King, W. W. The Downtonian and Dittonian strata of Great Britain and north-western Europe. *QJGS*, 90 : 526-70   75
[531] Read, H. H. Metamorphic geology of Unst in the Shetland Islands. *QJGS*, 90 : 637-88. (See also 553.)   55

1935

[532] Arkell, W. J. *The ammonites of the English Corallian beds.* London, Palaeontogr. Soc., 1935-48.   62, 71
[533] Davidson, C. F. The Tertiary geology of Raasay, Inner Hebrides. *Trans. Roy. Soc. Edinb.*, 58 : 375-407.   58
[534] Deer, W. A. The Cairnsmore of Carsphairn igneous complex. *QJGS*, 91 : 47-76.   57
[535] Dewey, H. *British regional geology: South-west England.* London, Geol. Surv. L; 3rd, 1969.   45, 50, 66, 72
[536] Eastwood, T. *British regional geology: Northern England.* London, Geol. Surv. L; 3rd, 1963.   66, 72, 73, 76
[537] Edmunds, F. H. *British regional geology: The Wealden district.* London, Geol. Surv. L; 4th, 1965.   50, 66, 82
[538] George, T. N. and Smith, B. *British regional geology: North Wales.* London, Geol. Surv. L; 3rd, 1961.   66, 78
[539] Pocock, R. W. and Whitehead, T. H. *British regional geology: The Welsh Borderland.* London, Geol. Surv. L; 2nd, 1948.
                                                                   46, 66, 84
[540] Pringle, J. *British regional geology: The south of Scotland.* London, Geol. Surv. L; 2nd, 1948.   57, 66
[541] Read, H. H. *British regional geology: The Grampian Highlands.* London, Geol. Surv. L; 3rd, 1966.   55, 66
[542] Read, H. H. The gabbros and associated xenolithic complexes of the Haddo House district, Aberdeenshire. *Trans. Roy. Soc. Edinb.*, 91 : 591-638. (See also 448, 481, 614, 615 and 631.)   55
[543] Richey, J. E. *British regional geology: Scotland: the Tertiary volcanic districts.* London, Geol. Surv. L; 3rd, 1961.
                                                                       58, 66
[544] Sherlock, R. L. *British regional geology: London and Thames valley.* London, Geol. Surv. L; 3rd, 1962.   50, 66
[545] Shotton, F. W. The stratigraphy and tectonics of the Cross Fell inlier. *QJGS*, 91 : 639-704.   70
[546] Swinnerton, H. H. The rocks below the Red Chalk of Lincolnshire, and their cephalopod faunas. *QJGS*, 91 : 1-46.   71

D

[547] Welch, F. B. A. and Crookall, R. *British regional geology: Bristol and Gloucester district.* London, Geol. Surv. L; 2nd, 1948.   *66, 76*

1936

[548] Chatwin, C. P. *British regional geology: The Hampshire basin and adjoining areas.* London, Geol. Surv. L; 3rd, 1960.   *26, 66*

[549] Edmunds, F. H. and Oakley, K. P. *British regional geology: The Central England district.* London, Geol. Surv. L; 3rd, 1969.   *66, 80*

[550] Macgregor, M. and MacGregor, A. G. *British regional geology: The Midland Valley of Scotland.* London, Geol. Surv. L; 2nd, 1948.   *56, 66*

[551] Matley, C. A. and Smith, B. The age of the Sarn granite. *QJGS*, 92 : 188-200. (See also 400, 488, 503, 517 and 563.)
                                                                    78

[552] Phemister, J. *British regional geology: Scotland: the Northern Highlands.* London, Geol. Surv. L; 3rd, 1960   *54, 55, 66*

[553] Read, H. H. Metamorphic correlation in the polymetamorphic rocks of the Valla Field block, Unst, Shetland Islands. *Trans. Roy. Soc. Edinb.*, 59 : 195-221. (See also 531.)   *55*

[554] Weir, J. and Leitch, D. The zonal distribution of the non-marine lamellibranchs in the Coal Measures of Scotland. *Trans. Roy. Soc. Edinb.*, 57 : 697-751.   *77*

[555] Wray, D. A. *British regional geology: The Pennines and adjacent areas.* London, Geol. Surv. L; 3rd,1954.   *66, 76, 77*

1937

[556] Anderson, J. G. C. The Etive granite complex. *QJGS*, 93 : 487-533.   *55*

[557] Chatwin, C. P. *British regional geology: East Anglia and adjoining areas.* London, Geol. Surv. L; 4th, 1961.   *66, 74*

[558] Jones, O. T. On the sliding or slumping of submarine sediments in Denbighshire, North Wales, during the Ludlow period. *QJGS*, 93 : 241-83. (See also 569.)   *78*

[559] Phillips, F. C. A Fabric study of some Moine Schists and associated rocks. *QJGS*, 93 : 581-620. (See also 578 and 611.)   *55*

[560] Pringle, J. and George, T. N. *British regional geology: South Wales.* London, Geol. Surv. L; 3rd, 1970.   *66, 76*

[561] Straw, S. H. The higher Ludlovian rocks of the Builth district. *QJGS*, 93 : 406-56.   **79**

1938

[562] Jones, O. T. The evolution of a geosyncline. *QJGS,* 94 : lx-cx.
67

[563] Matley, C. A. The geology of the country around Pwllheli,
Llanbedrog and Madryn, south-west Caernarvonshire.
QJGS, 94 : 555-606. (See also 400, 488, 503, 517 and 551.)
78

[564] Trueman, A. E. *The scenery of England and Wales.* London,
Gollancz. L. 53

[565] Wells, A. K. *Outline of historical geology.* London, Murby.
L; 5th, with J. F. Kirkaldy, 1966. 66, 76

1939

[566] Heard, A. Further notes on Lower Devonian plants from
South Wales. *QJGS,* 95 : 223-9. (See also 479.) 62

1940

[567] Allan, D. A. The geology of the Highland Border from Glen
Almond to Glen Artney. *Trans. Roy. Soc. Edinb.,* 60 :
171-93. (See also 485.) 75

[568] George, T. N. The structure of Gower. *QJGS,* 96 : 131-98.
(See also 478 and 525.) 72

[569] Jones, O. T. The geology of the Colwyn Bay district : a study
of submarine slumping during the Salopian period. *QJGS,*
95 : 335-82. (See also 558.) 78

1941

[570] Leitch, D. The upper Carboniferous rocks of Arran. *Trans.
Geol. Soc. Glasg.,* 20 : 141-54. 77

1942

[571] Anderson, E. M. *The dynamics of faulting and dyke formation.* Edinburgh, Oliver and Boyd. L. 55

1943

[572] Hudson, R. G. S. and Cotton, G. The Namurian of Alport
Dale, Derbyshire. *Proc. Yorks. Geol. Soc.,* 25 : 142-73.
(See also 577.) 77

1944

[573] Fearnsides, W. G. and Davies, W. The geology of Deudraeth :
the country between Traeth Mawr and Traeth Bach,
Merioneth. *QJGS,* 99 : 247-76. (See also 357 and 380.)
78

[574] Hollingworth, S. E., Taylor, J. H., and Kellaway, G. A. Large-scale superficial structures in the Northampton iron-stone field. *QJGS*, 100 : 1-44. (See also 579 and 617.)   *85*

[575] Holmes, A. *Principles of physical geology*. London, Nelson. L; 2nd, 1965.   *65*

1945

[576] Dunham, K. C. and Stubblefield, C. J. The stratigraphy, structure and mineralization of the Greenhow mining area, Yorkshire. *QJGS*, 100 : 209-68.   *76, 77*

[577] Hudson, R. G. S. and Cotton, G. The Lower Carboniferous in a boring at Alport, Derbyshire. *Proc. Yorks. Geol. Soc.*, 25 : 254-330. (See also 572.)   *76, 77*

[578] Phillips, F. C. The microfabric of the Moine Schists. *Geol. Mag.*, 82 : 205-20. (See also 559 and 611.)   *55*

1946

[579] Hollingworth, S. E. and Taylor, J. H. An outline of the geology of the Kettering district, and report of the Kettering field meeting. *Proc. Geol. Ass.*, 57 : 204-45. (See also 574 and 617.)   *85*

[580] Jones, O. T. and Pugh, W. J. The complex intrusion of Welfield, near Builth Wells, Radnorshire. *QJGS*, 102 : 157-88. (See also 589 595 and 603.)   *79*

[581] Kennedy, W. Q. The Great Glen fault. *QJGS*, 102 : 41-76.   *55*

[582] Lees, G. M. and Taitt, A. H. The geological results of the search for oilfields in Great Britain. *QJGS*, 101 : 255-317.   *82*

[583] Matley, C. A. and Wilson, T. S. The Harlech Dome, north of the Barmouth estuary. *QJGS*, 102 : 1-40.   *78*

[584] Steers, J. A. *The coastline of England and Wales*. Cambridge, University Press.   *53*

1947

[585] Anderson, J. G. C. The geology of the Highland Border: Stonehaven to Arran. *Trans. Roy. Soc. Edinb.*, 61 : 497-515.   *55*

[586] Arkell, W. J. *The geology of the country around Weymouth, Swanage, Corfe and Lulworth*. London, Geol. Surv.   *26, 71*

[587] Arkell, W. J. *The geology of Oxford*. Oxford, Clarendon Press.   *50, 71*

[588] Holmes, A. The construction of a geological time-scale. *Trans. Geol. Soc. Glasg.*, 21 : 117-52. (See also 387 and 639.)   *81*

[589] Jones, O. T. The geology of the Silurian rocks west and south of the Carneddau range, Radnorshire. *QJGS*, 103 : 1-36. (See also 580, 595 and 603.)   79
[590] Trotter, F. M. The structure of the Coal Measures in the Pontardawe-Ammanford area, South Wales. *QJGS*, 103 : 89-133.   77
[591] Trueman, A. E. Stratigraphical problems in the coalfields of Great Britain. *QJGS*, 103 : lxv-civ.   77
[592] Watts, W. W. *The geology of the ancient rocks of Charnwood Forest, Leicestershire.* Leicester, Leicester Literary and Philosophical Society.   72

1948

[593] Anderson, E. M. Lineation and petrofabric structure. *QJGS*, 104 : 99-132.   55
[594] Edwards, W. and Stubblefield, C. J. Marine bands and other faunal marker-horizons in relation to the sedimentary cycles of the Middle Coal Measures of Nottinghamshire and Derbyshire. *QJGS*, 103 : 209-60.   77
[595] Jones, O. T. and Pugh, W. J. A multi-layered dolerite complex of laccolithic form, near Llandrindod Wells, Radnorshire and the form and distribution of dolerite masses in the Builth-Llandrindod inlier. *QJGS*, 104 : 43-98. (See also 580, 589 and 603.)   79
[596] Kennedy, W. Q. The significance of thermal structure in the Scottish Highlands. *Geol. Mag.*, 85 : 229-34. (See also 604 and 610.)   55
[597] McLintock, W. F. P. (director). [Geological map of] *Great Britain* (Scale, 10 miles to 1 inch.) London, Geol. Surv. L.   68
[598] Wills, L. J. *The palaeogeography of the Midlands.* London, Hodder and Stoughton. L; 2nd, 1950.   67
[599] Wilson, V. *British Regional Geology: East Yorkshire and Lincolnshire.* London, Geol. Surv.   66, 71

1949

[600] Boswell, P. G. H. *The Middle Silurian rocks of North Wales.* London, Arnold.   78
[601] Eyles, V. A. and others. *The geology of Central Ayrshire.* London, Geol. Surv.   56
[602] Jones, O. T. The geology of the Llandovery district. The northern area. *QJGS*, 105 : 43-64. (See also 461.)   79
[603] Jones, O. T. and Pugh, W. J. An early Ordovician shore-line in Radnorshire, near Builth. *QJGS*, 105 : 65-99. (See also 580, 589 and 595.)   79

[604] Kennedy, W. Q. Zones of progressive metamorphism in the Moine Schists of the western Highlands of Scotland. *Geol. Mag.*, 86 : 43-56. (See also 596 and 610.)   55

[605] Taylor, J. H. *Petrology of the Northampton Sand ironstone formation*. London, Geol. Surv. (See also 609 and 649.)   85

[606] Trotter, F. M. The devolatilization of coal seams in South Wales. *QJGS*, 104 : 387-437.   77

1950

[607] Wright, J. *The British Carboniferous Crinoidea*. London, Palaeontogr. Soc., 1950-60.   62

1951

[608] Arkell, W. J. *English Bathonian ammonites*. London, Palaeontogr. Soc., 1951-8.   62

[609] Hollingworth, S. E. and Taylor, J. H. *The Northampton Sand ironstone: stratigraphy, structure and reserves*. London, Geol. Surv. (See also 605 and 649.)   85

[610] Kennedy, W. Q. Sedimentary differentiation as a factor in the Moine-Torridonian correlation. *Geol. Mag.*, 88 : 257-66. (See also 596 and 604.)   54

[611] Phillips, F. C. Apparent coincidences in the life-history of the Moine Schists. *Geol. Mag.*, 88 : 225-35. (See also 559 and 578.)   55

[612] Sutton, J. and Watson, J. V. The pre-Torridonian metamorphic history of the Loch Torridon and Scourie areas in the north-west Highlands. *QJGS*, 106 : 241-307.   54

[613] Wills, L. J. *A palaeogeographical atlas*. London, Blackie.   67

1952

[614] Read, H. H. Metamorphism and migmatization in the Ithan valley, Aberdeenshire. *Trans. Edinb. Geol. Soc.*, 15 : 265-79. (See also 448, 481, 542, 615 and 631.)   55

[615] Read, H. H. and Farquhar, O. C. The geology of the Arnage district : a re-interpretation. *QJGS*, 107 : 423-40. (See also 448, 481, 542, 614 and 631.)   55

[616] Whittard, W. F. A geology of Shropshire. *Proc. Geol. Ass.*, 63 : 143-297.   84

1953

[617] Kellaway, G. A. and Taylor, J. H. Early stages in the physiographic evolution of a portion of the East Midlands. *QJGS*, 108 : 343-75. (See also 574 and 579.)   85

[618] Sutton, J. and Watson, J. V. The supposed Lewisian inlier at Scardroy, central Ross-shire. *QJGS*, 108 : 99-126. (See also 623 and 640.) *54*

[619] Williams, A. The geology of the Llandeilo district, Carmarthenshire. *QJGS*, 108 : 177-208. *79*

1954

[620] Currie, E. D. Scottish Carboniferous goniatites. *Trans. Roy. Soc. Edinb.*, 62 : 527-602. *62*

[621] King, W. B. R. The geological history of the English Channel. *QJGS*, 110 : 77-101. *82*

[622] McIntyre, D. B. The Moine thrust : its discovery, age, and tectonic significance. *Proc. Geol. Ass.*, 65 : 203-19. *54*

[623] Sutton, J. and Watson, J. V. The structure and stratigraphical succession of the Moines of Fannich Forest and Strath Bran, Ross-shire. *QJGS*, 110 : 21-53. (See also 618 and 640.) *54*

[624] Trueman, A. E. (editor). *The coalfields of Great Britain.* London, Arnold. *77*

1955

[625] Anderson, J. G. C. The Moinian and Dalradian rocks between Glen Roy and the Monadliath Mountains. *Trans. Roy. Soc. Edinb.*, 63 : 15-36. *55*

[626] Bailey, E. B. Moine tectonics and metamorphism in Skye. *Trans. Edinb. Geol. Soc.*, 18 : 93-166. *54*

[627] Charlesworth, J. K. The later-glacial history of the Highlands and islands of Scotland. *Trans. Roy. Soc. Edinb.*, 62 : 769-928. *74*

[628] Jones, O. T. The geological evolution of Wales and the adjoining regions. *QJGS*, 111 : 323-51. *67*

[629] Kennedy, W. Q. The tectonics of the Morar anticline. *QJGS*, 110 : 357-82. *54*

[630] Whittard, W. F. *The Ordovician trilobites of the Shelve inlier, west Shropshire.* London, Palaeontogr. Soc., 1955-67. *62, 84*

1956

[631] Read, H. H. and Farquhar, O. C. The Buchan anticline of the Banff nappe of Dalradian rocks in north-east Scotland. *QJGS*, 112 : 131-56. (See also 448, 481, 542, 614 and 615.) *55*

[632] Sutton, J. and Watson, J. V. The Boyndie Bay syncline of the Dalradian of the Banffshire coast. *QJGS*, 112 : 103, 128. *55*

[633] Wills, L. J. *Concealed coalfields*. London, Blackie.   *47*

1957

[634] Charlesworth, J. K. *The Quaternary era*. London, Arnold.
                                                           74

1958

[635] Bulman, O. M. B. The sequence of graptolite faunas.*Palae-
     ontology*, 1 : 159-73.   *69*
[636] George, T. N. Lower Carboniferous palaeogeography of the
     British Isles. *Proc. Yorks. Geol. Soc.*, 31 : 227-318. *67, 76*
[637] Kennedy, W. Q. The tectonic evolution of the Midland
     Valley of Scotland. *Trans. Geol. Soc. Glasg.*, 23 : 107-33.
                                                           56

1959

[638] Allen, P. The Wealden environment : Anglo-Paris basin. *Phil.
     Trans. Roy. Soc.* (series B), 242 : 283-346.   *67*
[639] Holmes, A. A revised geological time-scale. *Trans. Edinb.
     Geol. Soc.*, 17 : 183-216. (See also 387 and 588.)   *81*
[640] Sutton, J. and Watson, J. V. Structures in the Caledonides
     between Loch Duich and Glenelg, north-west Highlands.
     QJGS, 114 : 231-8. (See also 618 and 623.)   *54*
[641] Williams, A. A structural history of the Girvan district, south-
     west Ayrshire. *Trans. Roy. Soc. Edinb.*, 63 : 629-67. *57*

1960

[642] Falcon, N. L. and Kent, P. E. *Geological results of petroleum
     exploration in Britain, 1945-1957*. London, Geol. Soc. *82*

1961

[643] Giletti, B. J., Moorbath, S., and Lambert, R. St J. A geo-
     chronological study of the metamorphic complexes of the
     Scottish Highlands. *QJGS*, 117 : 233-64.   *81*

1962

[644] Dearnley, R. An outline of the Lewisian complex of the Outer
     Hebrides in relation to that of the Scottish mainland.
     QJGS, 118 : 143-76. (See also 647.)   *54*
[645] George, T. N. Tectonics and palaeogeography in England.
     Sci. Prog., 50 : 192-217; 51 : 32-59. 1962-3.   *67*
[646] Williams, A. *The Barr and Lower Ardmillan series (Caradoc)
     of the Girvan district, south-west Ayrshire*. London, Geol.
     Soc.   *57*

1963

[647] Dearnley, R. The Lewisian complex of South Harris. *QJGS*, 119 : 243-307. (See also 644.)   *54*

[648] Johnson, M. R. W. and Stewart, F. H. (editors). *The British Caledonides*. Edinburgh.   *55, 57, 86*

[649] Taylor, J. H. *Geology of the country around Kettering. Corby and Oundle*. London, Geol. Surv. (See also 605 and 609.)   *85*

1964

[650] Harland, W. B., Smith, A. G., and Wilcock, B. (editors). *The Phanerozoic time-scale*. London, Geol. Soc.   *81, 86*

1965

[651] Craig, G. Y. (editor). *The geology of Scotland*. Edinburgh, Oliver and Boyd.   *54, 55, 56, 57, 58, 86*

1966

[652] Stubblefield, C. J. (director) and Dunning, F. W. *Tectonic map of Great Britain and Northern Ireland* (Scale, 25 miles to 1 inch.) London, Geol. Surv.   *68*

1967

[653] Rayner, D. H. *The stratigraphy of the British Isles*. Cambridge, University Press.   *66, 76*

[654] Harland, W. B. and others (editors). *The fossil record*. London, Geol. Soc.   *86*

1968

[655] Anderson, J. G. C. and Owen, T. R. *The Structure of the British Isles*. Oxford, Pergamon Press.   *66*

[656] Sylvester-Bradley, P. C. and Ford, T. D. (editors). *The geology of the East Midlands*. Leicester, Leicester University Press.   *86*

1969

[657] Bennison, G. M. and Wright, A. E. *The geological history of the British Isles*. London, Arnold.   *66, 67, 76*

[658] Kent, P. E., Satterthwaite, G. E., and Spencer, A. M. (editors). *Time and place in orogeny*. London, Geol. Soc.   *86*

[659] Wood, A. (editor). *The Pre-Cambrian and Lower Palaeozoic rocks of Wales*. Cardiff, University of Wales Press.   *86*

2

The Major Themes

NOTES

1 The themes have been taken as they have arisen during the survey of the progress of British geology from the earliest times to the present day. They do not keep within strict time-limits, but may be grouped as follows : numbers *1* to *18*, seventeenth and eighteenth centuries; *19* to *54*, chiefly concerned with the nineteenth century; *55* to *86*, chiefly concerned with the twentieth century.
2 Each theme is given a serial number so that references may be made in Section I and Appendix A.
3 The numbers in square brackets in the text refer to the items in Section I. Where an author is named and there is no number the reference is to Appendix A.

1 Allusions before the seventeenth century

The description of the geology of Britain can hardly be said to have been begun at any one definite time. A few very casual notices of facts that may be taken as 'geological' are to be found in British mediaeval writings; mention of the occurrence of, for instance, minerals, coal, and limestone.

We can start our list with the two well-known works of travel, topography, and antiquity : the *Itinerary* of John Leland, c. 1538 [1] and the *Britannia* of William Camden, [2]. The first remark we come across is Leland's highly significant one on stratification : that of the Trias in the Isle of Axholm, Lincolnshire. He says : 'The stones ly yn the ground lyke a smothe table : and be bedded one flake under another.' As Arkell and Tomkeieff remark (1953), the term 'bed' (thus 'bedded') is 'perhaps the commonest term in geology.' It is also the first term we meet in our journey through the annals of British geology. Among the other stray records made by these writers the most interesting are those of the occurrence of fossils. Leland records 'stones clerly fascioned lyke cokills, and myghty shells of oysters turned in to stones' from the Lower Oolites of Gloucestershire, and 'stones figured like serpentes' from the Lias at Keynsham, Somerset. Camden refers more particularly to the Keynsham fossils. These are obviously the ammonites, and the most conspicuous being the large well-known form *Arietites bucklandi*. Thus we can point to this fossil species as the first to have been definitely noticed in Britain.

Geology in general, before the seventeenth century—classical, mediaeval—is fully discussed by Geikie (1905), Adams (1938), and Bromehead (1945).

2 A formation mapped in words

George Owen's careful account, in manuscript, 1603 [3] of the course of the outcrop of the Carboniferous Limestone surrounding the South Wales coalfield makes a definite beginning to the deline- ation of the geology of Britain. Had he shown that he had a clear idea of a regional geological structure he would have been entitled to be called the first British geologist. At least it may be said that his account is the first attempt to map, if only in words, a British geological formation.

> 'You shall understand that the lymestone is a vayne of stones running his course for the most parte right east and west; although sometymes that same is found to wreath to the northe and southe. Of this lymestone there is founde two vaynes, the one smale and of noe great breadth [the north crop] the other verie broade [the south crop] . . . Between the said two vaynes there is a vayne if not severalle vaynes of coales. . . Whether I shall say that all the lande between these two vaynes should be stored with coales, I leave it to the judgemente of the skillfull myners, or those which with deepe judgement have entered into these hidden secreates.'

North (1931) and Bassett (1969(b)) give sketch-maps of this outcrop as now known, labelling particular parts with quotations from Owen's account.

3 Fossils described and their nature considered

It is in the second half of the seventeenth century, when there was a great outburst of scientific activity in Britain, that we find a real interest being taken in what was to be seen and found among the rocks, by far the most attention being given to fossils : 'fossils' in the old wide sense of any 'thing dug up', but especially 'fossils' in the modern sense of shells, bones, and other remains of organisms preserved in the rocks. Discoveries were then deliberately undertaken, the specimens were carefully described and figured, and vigorous discussion raged about their nature. They were difficult to account for because no correct geological notions were yet current and the origin of the rocks themselves, in which the fossils were found, was not inquired into. The fact that they were considered to be so mysterious probably stimulated their discovery. In the regional 'natural histories' of the time they took a remarkably prominent place. It was the riches of the English Jurassic rocks that supplied most of the specimens found.

'The earliest British writer to refer to fossils repeatedly was J. Childrey' (Cox, 1956). Childrey's 'observations', 1661 [4] have been brought to light and fully discussed by Smith (1942).

The work of Robert Hooke, 1665 [5] and 1705 [20] stands out from the rest for his detailed descriptions and drawings, particularly of ammonites, and his sound reasoning. From observations of their substance and occurrence he also discovered their true nature. Here is his description in his *Micrographia*, 1665 of ammonite shells, again from the Lias of Keynsham—certainly an historic fossil locality :

'Those serpentine or helical stones were cover'd or retained the shining or pearl-colour'd substance of the inside of a shel, which substance, on some parts of them, was exceeding thin. Some of them retain'd all along the surfaces of them very pretty kind of sutures which were most curiously shap'd in the manner of leaves, and every one of them in the same shell exactly one like another, which I was able to discover plainly enough with my naked ey but more perfectly and distinctly with my microscope. All these sutures, by breaking some of these stones, I found to be the termini or boundings of certain diaphragms or partitions which seem'd to divide the cavity of the shell into a multitude of very proportionate and regular cells or caverns. From all of which I cannot but think that all these, and most other kinds of stony bodies which are found thus strangely figured, do owe their formation and figuration to the shells of certain shel-fishes.'

Hooke's work has been reviewed by Gunther (1930), and Keynes (1960) has provided a bibliography.

Some seventy species of fossils are described and figured by Robert Plot in his work on *Oxfordshire*, 1677 [10]. Here we have excellent descriptions and beautifully engraved drawings of these objects from the Jurassic and Cretaceous. Among them are many well-known forms (given proper names by later authorities), such as the echinoid *Clypeus plotii*. He recognised the essential differences between those unrelated groups of bivalved shells, the brachiopods and the lamellibranchs. Similarly in his *Staffordshire*, 1686 [13] he describes and illustrates for the first time some of the now most familiar shells (brachiopods) from the Carboniferous and Silurian limestones (Challinor, 1945). Martin Lister, 1678 [11] brought to light further species and Edward Lhwyd, 1699 [17] wrote (in Latin) and illustrated the first work devoted entirely to British fossils. He described 1776 items and figured 267. Of these figures that of *Lithostrotion basaltiforme* is one of the most noteworthy. The following are other publications of the seventeenth century: William Somner, 1669 [6] described 'some strange bones' of *Rhinoceros antiquitatis* from river gravels in Kent (Edwards, 1967); John Beaumont, 1676 and 1683 [9] described Carboniferous crinoids ('rock-plants') from the Mendip Hills; Martin Lister, 1688 [14] added a few figures of fossils to a general work on shells; and Leigh, 1700 [19] described and illustrated some fossils from the Carboniferous Limestone of Derbyshire. For comments on Lister and his fossils see Jackson (1945).

Fossils, representing the chief Jurassic groups, occupy most of the large plates of illustrations in John Morton's *Northamptonshire*, 1712 [22]. Many are included in John Woodward's catalogue, 1728-9 [27] of a remarkably full and representative collection of fossils (in the old comprehensive sense), the collection itself in the original fine cabinets being still extant in the care of the Woodwardian Professor of Geology as the historical nucleus of the Sedgwick Museum, Cambridge. In 1728, also, Woodward [26] wrote a work which included 'brief directions for making observations and collections' of fossils, directions so complete and so much to the point that they put down in writing once and for all time the essentials of what the serious collector of geological specimens should do. In 1730 Samuel Dale [28] described the fossils from the highly fossiliferous Red Crag at Harwich.

In describing an ichthyosaurian skeleton from the Lias of Lincoln-shire William Stukeley, 1719 [24] wrote what appears to be the first paper in British geological literature devoted entirely to vertebrate palaeontology, and in 1758 W. Chapman and J. Wooller [32] des-cribed a crocodile from the Lias of Whitby.

Two important observations had been made which were in fact of fundamental importance as regards their geological and biological significance, though this was not realised by their discoverers: Lister, 1671 [7] noted that different kinds of rock yielded different fossils, and Lhwyd, 1695 [15] that fossils differed from the modern kinds of plants and shells.

Two illustrated fossil catalogues of the later part of the eighteenth century are rather isolated in time. D. C. Solander's descriptions and figures constitute the first fossil catalogue to be issued by the British Museum, 1766 [34]. It is a landmark in the history of research among British fossils. In it are described all the commoner fossils from that wonderfully prolific Eocene formation, the Barton Clay, with accurate and elegantly-drawn figures. The Linnaean specific nomenclature is adopted and the Latin language pervades the descriptions, but the whole style is thoroughly 'modern' and of the highest standard.

John Walcott's small book, 1779 [38] records and illustrates some of the fossils to be found in the neighbourhood of Bath.

4 Succession and structure

The first clear and detailed statement about a regional structure, at least for Britain, seems to be that given by George Sinclar in 1672 [8]. It is obviously a most important occurrence since the realisation that geology deals essentially with rock-structures in three dimensions is the first concept of the science. Sinclar explains the meaning of dip, strike, and outcrop ('dipp', 'streek', 'cropp') and, most important of all, describes folding and faulting. 'There remains only one question about the dipps and risings of coals which I shall a little consider, viz., whether coal and other metals, after they have declined such a length from their cropp, take another course and rise to the same point [level].' His consideration, which he embellishes with examples theoretical and actual, leads to a clear demonstra-tion of the synclinal structure of the Midlothian coalfield, and he goes on to particularise the effects of faults. There is also mention of igneous dykes and the effect of contact metamorphism ('these gaes that consist of whin-rock render the coal next to it as if it were already burnt'). Charles Leigh, 1700 [19] remarks on basin structure as being the common structure of coalfields ('strata shelving towards the center').

Robert Plot in his *Natural History of Staffordshire*, 1686 [13] gives a detailed stratigraphical succession through a part of the Coal Measures of the south of the country; Woodward, 1695 [16] makes important remarks on the structure of stratified masses and about the occurrence of fossils in them; and James Brewer, 1700 [18] carefully records the occurrence of fossil oyster-shells at a particular locality and in a particularised rock-sequence at the base of the Reading beds.

In a few words Thomas Robinson, 1709 [21] draws a contrast between the structures of the two mountainous regions of the northwest of England : between the 'easy depression' (slightly-dipping) Carboniferous Limestone strata composing 'the east mountains' and the 'quite different nature, consisting of a blue crag' of the Lower Palaeozoic 'western mountains.'

In two papers in the *Philosophical Transactions*, 1719, 1725, John Strachey [23] gives several versions of a geological section across the Somersetshire coalfield. A remarkable fact is that here, in what appears to be the first of such sections across any part of Britain, we have the portrayal of an unconformity, a phenomenon of the first importance but one not fully appreciated until Hutton explained it at the end of the century. We discuss unconformity under a later heading. Strachey's section is reproduced by Fitton (1832), Sheppard (1917), Mather and Mason (1939), and Tomkeieff (1962).

The title of John Michell's paper 'Conjectures [on] earthquakes' in the *Philosophical Transactions*, 1760 [33] does not indicate the great importance of what he has to say as to general geology. This is a discussion of the denudation of anticlinal regions resulting in duplicated parallel outcrops with the older rocks along the middle often forming mountainous tracts. He mentions vaguely 'an instance in Great Britain', where he may be referring to the Pennines. The fullest commentaries on this paper are those by Fitton (1832), Geikie (1918), and North (1931), and Mather and Mason (1939) give extensive extracts.

Imaginary and unnatural theories as to the formation and subsequent history of the earth were rife and voluminous from the latter part of the seventeenth to the latter part of the eighteenth centuries (theme *18*). John Whitehurst's 'inquiry', 1778 [37] into these matters is as practically worthless as any of the others but, as in the case of John Woodward, his book contains very valuable material. His 'observations on the strata in Derbyshire' really constitute a

quite separate treatise which is one of the main landmarks in the
progress of knowledge in British geology, being packed with significant facts and just inferences. He enunciates the principle of the
orderly superposition of strata, describing the succession of the
Carboniferous rocks of Derbyshire; and he discusses the character
of the fossils as affording evidence of marine or freshwater deposition.

> 'It may appear wonderful that amidst all the confusion of the
> strata there is nevertheless one constant invariable order in the
> arrangement of them and their various productions of animal,
> vegetable and mineral substances, or rather the figures or im-
> pressions of the two former. It would not be understood that
> the strata in every other part of the world are perfectly
> analogous to those in Derbyshire, or that their productions
> are the same : but that there is as much regularity in the
> arrangement of the strata in one country as there is in
> another.'

Whitehurst's other chief observations come under a later heading (theme *8*). The present writer has made a commentary
(1947).

John Williams, in his *Natural History of the mineral kingdom*,
1789 [43] gives us penetrating observations on a wide variety of
geological topics, his descriptions of faulting and folding being
extraordinarily complete. 'As there is nothing so effectual as local
examples to enable us to form proper ideas, let us go to the rocky
shores of the ocean, or to some river which has cut deep into the
rock, and chuse out a fair and lofty section of the strata.' In particular he is enthusiastically insistent that the 'noble section' of
Carboniferous rocks at Gilmerton, near Edinburgh, should be seen,
and the folding of the rocks into 'a kind of trough' (an elongated
basin) appreciated. He also explains how the structure and succession is proved by horizontal levels and vertical shafts.

5 Some miscellaneous observations

Observations of various kinds recorded between 1695 and 1778
may be summarised as follows :

Lhwyd, 1695 [15]. Columnar jointing in the (igneous) rocks of
the Glyder Mountain, North Wales : 'The hill is adorned with
numerous equidistant pillars and these again slightly cross'd at
certain joints.' Fossils of plants from the Coal Measures of North

E

and South Wales (figures show *Sigillaria, Pecopteris,* and *Neuropteris*)

Leigh, 1700 [19]. Rock-salt in Cheshire and lead ore in Derbyshire

Packe, 1743 [29]. Topographical map of East Kent, with a commentary, incorporating stray geological facts

Owen, 1754 [30]. Remarks on 'earths, rocks, stones, and minerals', including ammonites ('snake-stones') of the Bristol neighbourhood (See Smith, 1945)

Borlase, 1758 [31]. Remarks on the rocks and minerals of Cornwall

Papers in the *Philosophical Transactions,* 1754-64. Seven papers on particular groups of fossils. These are not included separately in our list. The two most palaeontologically 'advanced' are those on belemnites by Gustavus Brander, 1754, and Joshua Platt, 1764.

Banks, 1774 [35]. The columnar basalt of Staffa fully described

Pryce, 1778 [36]. The geological setting of mineral lodes

6 The first tables of strata

In addition to their other features the two papers by Strachey, 1719-25 [23] are of great interest in that they are the first to give us a stratigraphical table of British formations. The following succession is recognised (in modern terminology) : mineral-bearing rocks below the Coal Measures, Coal Measures, New Red Sandstone, Lias, Oolite, Chalk. The next table we have is the one jotted down by John Michell on part of the cover of a letter bearing the post-mark November 21st, 1788, found and published by John Farey in 1810 [41]. This adds several formations to Strachey's list : Magnesian Limestone ('Roch Abbey and Brotherton limes'), Bunter ('Sherewood Forest pebbles and gravel'), Lower Greensand ('Sand of Bedfordshire'), Gault ('Golt').

7 The mapping of outcrops

We have seen (theme 2) that, so far as is known, the first piece of geological investigation in Britain was the tracing of an outcrop by Owen, 1603 [3] and this is particularly appropriate since that must always be the main part of the geological exploration of any territory large or small.

Martin Lister, 1684 [12] seems to have been the first, at least in Britain, to propose the making of geological maps ('soile or mineral maps, as I may call them'). He suggests how Yorkshire and Nottinghamshire might be shown as parcelled out on such a map. Although he dimly realises that 'something more might be comprehended from the whole and from every part than I can possibly foresee', he had no correct notions of stratification and structure.

In 1723 Benjamin Holloway [25] traced the outcrops of the Lower Greensand and Chalk strata forming parallel ridges, in Bedfordshire and neighbouring counties, and very significantly, realised that these rock-masses were 'disposed in the earth like strata, coming through vast tracts.' In 1782 John Strange [39] followed the outcrop of the Lias with its *Gryphaea* ('Gryphites oyster) fossils, across England and Wales. We have here for the first time the realisation of the use of fossils in following the continuation of a stratigraphical formation across country. This same formation, the Lias, was again traced, and apparently independently, across England and Wales by John Smeaton, 1791 [45]. Further manuscript material of Michell's, rescued by Sir Archibald Geikie, is of correspondence with Henry Cavendish on the Mesozoic rocks of north-east England, 1788 [42] in which that most important stratigraphical-structural condition of overstep is suggested, surely for the first time, and correctly applied to the base of the Chalk. Cavendish writes:

'I believe you must be right in supposing your yellow limestone [Permian Magnesian Limestone] to be quite distinct from the other. From what I can learn, I believe the N.W. edge of the other [Jurassic Lower Oolites], after running from Gunnerby Hill on the E. side of the Trent, crosses the Humber and runs under the Yorkshire Chalk, and appears again about Castle Howard, and so runs to Scarborough, the Chalk in that place lapping over and extending further west than the limestone.'

8 Basalt: dykes and sills: contact metamorphism

We have seen (theme *4*) that Sinclar [8], as far back as 1672, had noticed contact metamorphism by igneous dykes in the Midlothian coalfield. Whitehurst, in his account of the Carboniferous strata of Derbyshire, 1788 [37] carefully considered the question of the origin of the intercalated igneous rocks (basalts and dolerites), there called 'toadstones', and in doing so made what is perhaps his most significant contribution to geology. This question of the mode of origin of these rocks was in effect the question of the origin of all rocks of a like kind.

Desmarest in France had demonstrated in 1774 the existence of volcanic activity in past geological ages and the presence of igneous rocks in the crust of the earth; Whitehurst arrived at the same conclusion four years later, apparently quite independently and by direct inference from the examination of a rock which, in the mass, behaved structurally like a sedimentary stratum and which was situated in a region, indeed in a whole country, showing now not the slightest signs of any recent volcanic activity. He was confronted with a rock which in a hand-specimen was so different from the limestone with which it was associated and so similar to specimens he had seen of recent lavas, that he had no hesitation in assigning to it a volcanic and not a sedimentary origin. Whitehurst considered all the toadstones to be intrusive :

'Toadstone is actual lava, and flowed from a volcano whose funnel, or shaft, did not approach the open air, but disgorged its fiery contents between the strata in all directions. Another remarkable phenomenon accompanying the Derbyshire lava is that the stratum of clay lying under No. 6 is apparently burnt, as much as an earthen pot or brick.'

While the basalts were probably sub-aerial lava-flows, the dolerites were certainly intrusive, producing contact metamorphism above and below. These toadstones were described by Bemrose, 1907 [365] and in the discussion of that paper Archibald Geikie referred to Whitehurst's observations. (See Challinor, 1947.)

Hutton, in his 1788 paper [40], gives a particularised account of the concordant and discordant minor intrusions (dykes and sills) of 'whinstone', though he may not always have distinguished between them and the contemporaneous lava flows. Playfair, 1802 [54] gives even more detailed accounts, emphasising the accompanying metamorphism. Hutton and Playfair were not only the foremost 'plutonists' (as regards the origin of major intrusions, granite in particular) but also among the foremost 'vulcanists', advocating the view that basalt and dolerite, which they tended not to discriminate, were formed by volcanic or associated activity.

Faujas Saint Fond is well-known as being one of the early demonstrators of the volcanic origin of basaltic rocks, particularly in his *Recherches sur les volcans éteints du Vivarais et Velay* (1778). In the tour of Britain which he made in 1784 [48] (but did not publish till 1797 owing to the troubles of the French revolution) he had no difficulty in recognising the basalts and other extrusive and minor intrusive rocks of the Glasgow and Edinburgh districts, the Oban district, and the Inner Hebrides (Staffa and Mull), and assumes their volcanic nature without argument, but he had no idea that they were of widely different geological ages.

Jameson, 1800 [52] in describing dykes in Arran, notes the composite nature of some of them.

9 Modern geology founded by the formulation of a reasonable 'theory of the Earth'

In 1788 British geology, as indeed geology in general, made a great stride forward with the publication in the first volume of the *Transactions of the Royal Society of Edinburgh* of the paper on the 'Theory of the Earth' by James Hutton [40]. (An abstract had been published in 1783.) Here, at last, after a few adumbrations on the European continent, we had a reasonable theory, destined to become the foundation of modern geology. It firmly established the conception of the geological cycle and insisted on the length of geological time : time stretching, without imaginable limit, back into the past and forward into the future. This carried with it the implication of the doctrine of uniformitarianism : an essential uniformity of process, varying in kind and degree from place to place and from time to time but with an over-all smooth continuity without world-wide catastrophes interrupting the normal course of nature. A very important feature of his theory was that granitic rocks were of igneous, not aqueous, origin. This opinion was opposed by the famous German mineralogist, Werner, and his powerful disciples. Hutton's views soon prevailed and are now accepted as obvious. Unfortunately Hutton pushed too far his advocacy of the effect of heat, and he wrongly supposed that it was the means of the consolidation of strata.

Hutton's 1788 paper was reprinted as the first chapter of his famous two-volume book of 1795 [47]. The greater part of the rest of the book deals also with theoretical matters. All these are notoriously tedious to read, not so much because of the occasional obscurity of particular passages but because of the author's excessive prolixity.

Hutton's theory has been extensively examined and criticised from the time it was first put forward, and it has been subjected to much scholarly commentary. We may cite particularly the following : Geikie (1871 and 1905), Bailey (1950), V. A. Eyles (1950), Tomkeieff (1950), V. A. Eyles and J. M. Eyles (1951), Bailey (1967), Dott (1969).

There is however much factual material embodied in his theoretical discussion, some of it in his original paper but mostly in the 1795 book, and his great contributions to the building up of our knowledge about the geological structure of Britain have been comparatively neglected. In perusing these volumes the reader is relieved to come across Hutton's graphic descriptions of particulars with the direct generalisations he makes from them, as they are admirably graphic and to the point. These particulars are separately commented on under the appropriate themes which follow in this section.

In 1802 was published John Playfair's *Illustrations of the Huttonian Theory of the Earth* [54]. It is a well-known masterpiece of geological literature, clarifying Hutton's writings in direct and elegant language and adducing many additional 'illustrations', that is examples of actual occurrences, from his own observations.

'For precision of statement and felicity of language it has no superior in English scientific literature. To its early inspiration I owe a debt which I can never fully repay. How different would geological literature be to-day if men had tried to think and write like Playfair!' (Geikie, an author of even greater literary masterpieces than Playfair's, in *Founders of Geology*, 1905. See also Geikie, 1906).

10 Rocks and structures in the Highlands of Scotland

John Williams's *Natural History of the mineral kingdom,* 1789 [43] is a mixture of observation, generalisation, and speculation. The speculative part contains good reasoning mixed with what now appears to be absurdity. In his preface he criticises Hutton's 'theory' which, in its original form [40], had appeared after he had finished writing his book. In the main Williams's criticisms are ill-founded, but on the consolidation of strata he is right where Hutton is wrong, and if Hutton had adopted the same simple ideas as Williams did the Huttonian Theory would have been without its one serious blemish.

Williams's observations, though chiefly on the geology of Scotland, range widely, and there is so much that is sound in the generalisations and principles that he derives from them that the work might perhaps be said to constitute the first textbook of geology (see theme 4). The two volumes run to over one thousand pages and the author, like Hutton, goes to quite unnecessary length in telling us what he has to say. We quote a few of his more significant remarks on Scotland under the several headings.

Coal measures occupy limited areas and being separated by under-lying rocks are not now connected between one coalfield and another

'The result of my enquiries about the Lothian coal fields establishes two important points in the natural history of coal. First, that every coal field is of limited extent and dimensions and that the coals crop out to the surface of the ground every where quite round the coal field; and, therefore, secondly, the coals do not stretch away under our high mountains and emerge upon the other side of them.'

Banded gneiss in Scotland

'There is a curious variety of stone found in the Highlands, which may be called streaked or stripped . . . commonly white and black alternately.'

Cambrian quartzite of the north-west Highlands

'The regularly stratified quartzy mountain rock. . . . There are large and high mountains of this stone in the shires of Ross and Inverness, which on a clear day appear at a distance as white as snow.'

The Old Red Sandstone—see theme 11
The structure of Ben Nevis, granite below and bedded (volcanic) rocks in the upper part

'The granite or porphyry of this mountain is all as one mass . . . but notwithstanding the uniformity of the basis and the great bulk of this mountain, the summit of it is, nevertheless, regularly stratified with a different stone.'

The structure and character of the Dalradian series of the Grampian highlands

'I have in that country traced a particular class of strata for near two hundred miles upon the bearing, which is nearly from north-east to south-west. . . . The strata of these lofty mountains which I have observed to be so very regular, generally decline with an easy slope towards the south-east. . . . There are a great variety of strata . . . such as several species of argillaceous strata, of schistus, and of slate, several species of limestone, and some marble. Also the micaceous mountain rock and the regularly stratified, hard, granulated, sugar-loaf-stone, composed chiefly of minute quartzy grains.'

11 *The Old Red Sandstone of Scotland*

Williams, 1789 [43], Hutton, 1795 [47], Faujas Saint Fond, 1797 [48], Jameson, 1800 [52], and Playfair, 1802 [54] each make significant remarks about the Old Red Sandstone of Scotland. These are best given, as selections, in their own words.

Williams

'The grey flaggy strata of Caithness, so thin and regular, and raised so light and broad, that three or four of them cover the side of a small house.'

Hutton

'We were now well acquainted with the pudding-stone [a basal conglomerate] which is interposed between the horizontal and alpine strata. In the island of Arran there is also a pudding-stone, even in some of the summits of the island. In like manner to the north of the Grampians, along the south side of Loch Ness [and] also found upon the south side of those mountains in the shire of Angus.'

Faujas Saint Fond (translation of 1799)

'The traveller sees with astonishment, in the environs of Oban, vast walls of pudding-stone which extend uninterruptedly along the coast. The different stones which enter into the composition of the pudding-stones are white quartzes, greenish trapp, argillaceous schistus, black calcareous stone, porphyries and compact, black, basaltic lava. All these different stones are thrown together, and intermingled without order, and agglutinated with a cement so hard, that it is exceedingly difficult to separate them with a hammer, which in general rather breaks than disjoins them.'

Jameson

'Wick is the chief county town in Caithness. The strata in the neighbourhood of the town are composed of sandstone flag, which is disposed in strata nearly horizontal.'

Playfair

'I shall here mention another mark of violent fracture that has been observed in rocks of breccia or pudding-stone. In rocks of this kind it sometimes happens that considerable portions are separated from one another, as if by a mathematical plane, which had cut right across all the quartzy pebbles in this way. Lord Webb Seymour and I observed pudding-stone rocks, exhibiting instances of this singular kind of fracture, near Oban, in Argyleshire, about three years ago.'

12 *The igneous origin of granite*

The origin of granite as a process of crystallisation and firm consolidation from a molten state, emplaced by intrusion into preexisting rocks, is one of the main points of the Huttonian theory. Hutton's supporters in this connection came to be known as 'plutonists'. In his 1788 paper [40] Hutton describes a piece of graphic granite from Portsoy, Aberdeenshire, as evidence that the component 'sparry and siliceous substances had been mixed together in a fluid state'. The matter is more fully considered in the separate papers by Hutton and James Hall in 1790 [44]; but the fullest discussion, with the lively account of the discovery of granite penetrating the strata in Glen Tilt, Perthshire, is to be found in the third volume of the 1795 book [47], rescued by Archibald Geikie and published in 1899. The following is an extract :

'On the south side of the glen, the strata are composed of alpine schistus, particularly of granulated quartz and micaceous limestone; and these strata dip into the hill in descending to the south. On the other side of the glen, the steep face of the hill is all covered with lumps of beautiful red granite, not a particle of which is to be seen upon the south side. Here therefore we are in the very spot which we desired, and fortunately for our researches, the river lays bare enough of the solid parts to give the most satisfactory view of what had been transacted in a former period, probably at the time the strata, which were originally horizontal, had been broken and displaced. It must be recollected that the present question regards the granite, how far it is to be considered as a primary mass in relation to the alpine schistus. I had in a preceding part of this work drawn a very probable conclusion concerning the natural history of granite, so far as those masses might be considered as analogous to basaltes, or subterraneous lava, in having been made to flow. We have both those points now perfectly decided; the granite is here found breaking and displacing the strata in every conceivable manner, and interjected in every possible direction among the strata which appear. This is to be seen, not in one place only of the valley, but in many places, where the rocks appear, or where the river has laid bare the strata. The granite must be considered as posterior to those strata, notwithstanding that these are found superincumbent on the granite.'

Playfair, in his life of Hutton (1805), tells us :

'The sight of objects which verified at once so many important conclusions in his system, filled him with delight; and as his feelings, on such occasions, were always strongly expressed, the guides who accompanied him were convinced that it must be nothing less than the discovery of a vein of silver or gold that would call forth such strong marks of joy and exultation.'

13 Unconformity sought, found, and described

Strachey's startling depiction, 1719 [23] of an unconformity (theme 4) was fortuitous. Hutton and Playfair realised that if they could find clear cases of structural unconformable relationship, representing a considerable time-gap, between two groups of stratified rocks, this would be the surest evidence they could get to confirm the validity of Hutton's main thesis, what we now usually call the 'geological cycle' or 'geostrophic cycle'.

Hutton, 1795 [47] found three unconformities in southern Scotland. The first was that in the Isle of Arran (Lower Carboniferous on Dalradian), where an unconformity is obvious but the actual plane deceptive; the second (Old Red Sandstone on Silurian) was in the valley of the Jed, now overgrown and difficult to find; the third was at Siccar Point on the Berwickshire coast, also with Old Red Sandstone resting on Silurian. This unconformity is still as striking as ever, kept fresh by the action of the sea, and it is certainly the most famous single exposure of a British unconformity. Part of Hutton's description of the finding of this unconformity, which was made by boat in company with Sir James Hall and John Playfair, is as follows :

'Dunglass burn is almost the boundary between the vertical [Silurian] and horizontal [Old Red Sandstone and Carboniferous] strata. To the north-west of this burn and beautiful dean are situated the coal, lime-stone, marsh, and sand-stone strata . . . they approach the schistus of which the hills of Lammermuir to the south are composed. Though the boundary between the two things here in question be easily perceivable from the nature of the country at the first inspection by the rising of the hills, yet this does not lead one precisely to the junction; and in the extensive common boundary of those two things, the junction itself is only to be perceived in few places, where the rock is washed bare by the rivers or the sea, and where this junction is exposed naked to our view. The sea is here wearing away the coast. . . . The solid strata are everywhere exposed either in the cliff or on the shore; we were therefore certain of meeting with the junction in going from Dunglass to Fast Castle, which is upon the schistus. But this journey can only be made by sea . . . Having taken boat at Dunglass burn, we set out to explore the coast. At Siccar Point, we found a beautiful picture of this junction washed bare by the sea. The sand-stone strata are partly washed away, and partly remaining upon the ends of the vertical schistus; and, in many places, points of the schistus strata are seen standing up through among the sand-stone, the greatest part of which is washed away. Behind this again we have a natural section of those sand-stone strata, containing fragments of the schistus.'

Playfair, 1802 [54] sought other unconformities. He found several, among them two exposures of the very pronounced unconformity between Lower Carboniferous and Lower Palaeozoic in the Ingleborough district.

Jameson, 1805 [57] by a diagram of another Scottish example explained this structural relationship, using in doing so the term 'conformable', apparently for the first time in a geological sense.

All the early examples of unconformity are described, pictured, and discussed in Tomkeieff's admirable historical study (1962).

14 Lower Palaeozoic fossils

Just as unconformity was 'needed' to confirm Hutton's theory of the geological cycle, so was it necessary to prove that the so-called 'primitive' rocks had an origin no different from that of the so-called 'secondary' rocks, all being formed by the same timeless processes. Thus the finding of fossils in the 'primitive' rocks by Hutton and his several companions was an exciting and momentous discovery for him and his theory. He records such fossils from Wales, the Lake District, and southern Scotland; his account of his first find, from the Ordovician Coniston Limestone, is as follows, 1795 [47]:

> 'In this summer 1788, coming from the Isle of Man, Mr. Clerk and I travelled through the alpine schistus ['primitive' rock] country of Cumberland and Westmorland. We found a lime-stone quarry upon the banks of Windermere, near the Low-wood Inn. I examined this limestone closely, but despaired of finding any vestige of organised body. Fortunately, however, I at last found a fragment in which I thought to perceive the works of organised bodies in a sparry state. I have brought home this specimen, which I have now ground and polished; and now it is most evidently full of fragments of entrochi [crinoids]. Here is one specimen which at once overturns all the speculations formed upon that negative proposition [that the 'primitive' rocks are without fossils]. The schistus mountains of Cumberland were, in this respect, as perfect primitive mountains as any upon the earth, before this observation; now they have no claim upon that score, no more than any lime-stone formed of shells.'

Playfair, 1802 [54] adds to these northern finds by recording fossils from the Devonian limestone at Plymouth.

15 Miscellaneous observations in Scotland

THE FORMER EXTENSION OF THE BRITISH ISLES
It is a clear inference that the present outline of the British Isles must be more recent, in any of its parts, than the formation of the most recent rocks seen to be cut through but which are otherwise continuous from one island to another, or from Britain to the Continent. This seems to have been first definitely put into words by Williams and Hutton. Their statements are as follows:

> Williams [43]—'When I was upon the summit of Beneves, it appeared to me evident that the Hebrides or western isles of Scotland had been at first formed all in one continuous stretch of land with the Highlands and with one another.'
> Hutton is much more specific [47]—'The east coast of Caithness is a perpendicular cliff of sandstone, lying in a horizontal position, and thus forming a flat country above the

shore. But along this coast there are small islands, pillars, and peninsulas, of the same strata, corresponding perfectly with that which forms the greater mass. Now, shall we suppose those strata of sandstone to have been formed in their place, and to have reached no farther eastward into the sea?—It is impossible . . . . In following this connection of things, we cannot refuse to acknowledge that Ireland had formerly been in one mass of land with Britain, in like manner as the Orkneys had been with Scotland. It will be still less possible to refuse the junction of England with the continent of France; the testimony of that peculiar body of chalk and flint, which borders each of those opposite coasts, forms an argument which is irrefragable.'

GLACIAL DRIFT AND RAISED BEACHES
The observations are again those of Williams and Hutton.

Williams—'Every observing person must have frequently seen in plains, at a little distance from large and high mountains, a prodigious quantity of coarser and finer gravel, some of it containing large boulders. . . . This is often spread out with a plane superficies, and as often accumulated into long, round, oval, or semi-circular hillocks [drumlins].'
Hutton—'In the Firth of Forth there are, in certain places, particularly about Newhaven, the most perfect evidence of a sea bank, where the washing of the sea had worn the land, upon a higher level than the present . . . above Kinneel, there is a bed of oyster shells some feet deep appearing in the side of the bank, about 20 or 30 feet above the level of the sea, which corresponds with the old sea banks. There are many other marks of a sea beach upon a higher level than the present.'

CLEAR EVIDENCE OF SEDIMENTATION AND SUBSEQUENT FOLDING
The most momentous geological expedition ever made, by land or water, was perhaps Hutton and Hall's boat-trip along the Berwickshire coast. They not only made the famous discovery of the unconformity at Siccar Point (theme *13*), but Hutton relates how the key-process in geological history (and in *his* geological history) was irrefutably plain to see in the Silurian rocks of the sea-cliffs—the original process of the laying down of sediments in all respects similar to those being deposited in the shallow seas of today, and the subsequent deformation of these consolidated sediments. One of the pictures in the first volume of the *Theory of the Earth* represents

one of those examples; 'it was drawn by Sir James Hall from a perfect section in the perpendicular cliff at Lumisden burn. Here is a fine example of the bendings of the strata.' (An acute anticline and syncline are shown.) The voyagers then go on to describe the andesitic lava (of Old Red Sandstone age) of St. Abbs Head. Finally Hutton remarks: 'Here ended our expedition by water'. Certainly a good day's work!

DIFFERENTIAL EROSION : AND THE FORMS OF HILLS ARE DEPENDENT ON THEIR GEOLOGICAL STRUCTURE

Probably the most important principle in geomorphology, and the most obvious, is that the harder rocks resist erosion more effectively than the softer. Hutton gives much consideration to this matter and exemplifies it by giving a sketch of the physical features of Scotland as correlated with the occurrence of the broad regions of the harder and softer rocks. The same effect is produced round the coasts : 'The sea has left them in a shape corresponding to the composition of the land, in destroying the softer, and leaving the harder parts.' Neither Hutton nor later writers have implied that this is the only factor determining the heights of regions or the forms of coast; the relative importance of the various factors is still very much a matter of debate. Hutton also shows that the forms of individual hills, groups of hills, and ridges are dependent on their rock-composition and geological structure; and in particular he gives us the essential condition of the most usual kind of escarpment: 'an inclined rock forms a mountain sloping on one side, and having a precipice upon the upper part of the other side, with a slope of fallen earth at the bottom.'

DISCOVERY OF MESOZOIC ROCKS IN SCOTLAND

In the course of Faujas Saint Fond's journey through Scotland [48], this French 'vulcanist' discovered Liassic limestones with belemnites, beneath Tertiary lava on the east coast of Mull. His account of the discovery is as follows (in the translation of 1799) :

'About half a mile from Achnacregs the waves have brought to light a bed of lime-stone, that formerly lay buried under a current of black basaltic lava, of which the whole coast is formed. This bed is completely uncovered, and loses itself in the mass of lavas which rise into hills as they recede from the coast. The lime-stone is grey, hard, and brittle. I found some belemnites in it, the largest of which were five inches in length, and an inch and a half circumference towards the base.'

His mention of the belemnites fixes the Mesozoic age of the limestone now known to be Lias.

16 The western counties of England

In William George Maton's *Western counties of England*, 1797
[49] we have the first 'regional geology' of any large part of the
country. It is a topographical account of two journeys, one made in
1794 and the other in 1796, described in two corresponding volumes.
These accounts include a wealth of geological observations recorded
as they were met with in the course of the journeys. Thus the
geological material is scattered. His attempt to organise it as a
whole in a geological map was not successful, as the map is very
sketchy and inaccurate and indeed does not do justice to his statements which often contradict it. It is however of considerable
historical importance that the attempt should have been made. His
classification of the rock-groups is purely lithological. We can sort
them out and put them in stratigraphical order, but he gives us no
stratigraphical succession himself.

We here summarise some of his remarks, arranged in a geological
order.

SERPENTINE

The serpentine of the Lizard (Kynance Cove). The variations in
colour and the 'greasiness to the touch' are described, with mention
of some of the minerals ('chlorite', 'asbestos', 'steatites').

GRANITE

At several places, in particular Mount's Bay, Land's End, St Kitt's
Hill. The most precise mineralogical description is that of loose
boulders at Ivybridge, brought down by the streams from the near-
by outcrop on the north : 'a dead whitish colour, and composed of
a very large proportion of fel spar (which appears for the most part
in long narrow crystals), pellucid quartz, some schoel, and a few
scarcely discernible specks of mica.'

DEVONIAN

The limestone about Torquay, 'takes so good a polish as to obtain
the denomination of marble'; the 'argillaceous slates' and 'killas' of
North and South Cornwall, with remarks on the 'excellent slate for

roofs' of Delabole ('Denyball'); the 'coarse kind of compound grit-stone' of the Quantock Hills and Exmoor.

CARBONIFEROUS LIMESTONE
Brief mention of the occurrence in Somerset, and the resemblance between the limestone of the Mendip Hills and that of the Derby-shire Peak District. There is a beautiful wash-drawing of the Cheddar gorge.

CULM MEASURES
An anticline on the coast near Hartland

PERMO-TRIAS
The Thorverton basalt of the Exeter volcanic series: 'exhibits a cellular appearance like toad-stone [Derbyshire] . . . a lava-like texture . . . the Vulcanist perhaps will pronounce it the effect of fusion, but the cautious mineralogist will set it down among amygdaloidal earths without venturing to speculate upon its origin.' The cliffs of 'a deep red colour' at Teignmouth and the 'red loam soil' of the Vale of Taunton.

LIAS
Maton's remarks on the Lias are more telling and more geologically interesting than his more casual remarks on the other rocks he came across. The cliffs at Lyme Regis

'abound with madreporae [corals etc], ammonitae, belemnitae, and skeletons of fishes and other animals in a fossil state. All curious productions are diligently collected by a man living at Charmouth who is generally known throughout the country by the name of the Curi-man. There is a good deal of pyrites and bituminous matter in the soil, which has often taken fire after heavy rains.' The Ammonite Marble (Lower Lias) of Marston Magna is 'a very uncommon and beautiful species of stone, containing a congeries of animonitae, the nacre of the shells being still visible, the shells being filled with variegated sparry matter . . . when polished the whole becomes an elegant, ornamental substance for chimney-pieces, etc.'

The Ham Hill Stone (Upper Lias) is used 'in many buildings' which 'though so soft when first taken out of the quarry becomes hardened by the weather to an extraordinary degree.' The Blue Lias (Lower Lias) and White Lias (Rhaetic) are discriminated and compared. Maton suggests that the limestone bands in the Lias may be due to a secondary stratification : 'calcareous matter seems to have oozed, as it were, from the marl'. This is still a question today. Finally we again come across the Keynsham ammonites (Lower Lias) of the early reporters :

'We find immense *cornua ammonis*, which are carefully picked out and polished for sale by the quarrymen, who give them the appellation of snakestones. The diameter of many of these extraordinary fossils measures nearly two feet.'

KIMMERIDGE, PORTLAND, AND PURBECK
The Kimmeridge 'coal', and the Portland and Purbeck beds of Purbeck, Lulworth, and Portland are mentioned.

CHALK
An attempt to map the extent of the Chalk outcrop, 'wishing to ascertain as accurately as possible the course of the chalk in this county [Dorset], and the appearance of its boundaries.'

BOVEY TRACEY BEDS (OLIGOCENE)
A somewhat detailed description of the clays and lignites. 'The coal retains the vegetable structure and has exactly the appearance of charred wood.'

Maton refers to the correspondence between plant distribution and the nature of the underlying rock :

'Before we searched into the nature of the subsoil [south of Sherborne] we were sufficiently instructed that it has passed into chalk by the altered aspect of vegetation. By attending to this circumstance, the mineralogist may often obtain pretty certain indications of the transitions of strata. The botanist will afford information to the mineralogist. He will often tell him, by the preference of one particular plant, such, for instance, as *Hedysarum onobrychis* (saint foin), *Campanula glomerata,* or even the humble little *Hippocrepis comosa* (horse-shoe vetch) that the soil can be no other than a cretaceous one.'

Although Maton's geological notes are very scrappy it will be seen that they cover a wide field and are often much to the point. Nothing had been done previously as to the geology of this wide area, except for some mineralogical and practical notes on the Cornish mining district.

Maton's observations, especially as they were of such a pioneering nature, have been undeservedly neglected; Conybeare and De la Beche, in the eighteen-twenties and eighteen-thirties just refer to

F

the work, but it has since remained in almost total oblivion. Butcher (1968) refers to it briefly. Two previous remarks on parts of the district should however be mentioned: Hutton, 1788 [40] on the outcrop of the Chalk in the Isle of Wight and its continuation (re-appearance) in Dorsetshire, and Williams, 1789 [43] on the jointing in the Cornish granite.

17 Knowledge at the end of the eighteenth century

By the end of the eighteenth century the rocks of Britain had been classified into two main groups, the Primary (at first called 'Primitive') and the Secondary. This grouping was based on observed superposition, particular attention being paid to uncon-formities, and as the names show, on the consequently inferred relative age. There is in fact a natural two-fold grouping in many regions, but the stratigraphical (time) gap is not everywhere the same; for instance, over much of Scotland it is below the Old Red Sandstone, in north-west England below the Carboniferous, while in south Wales and south-west England it is below the Permo-Trias or the Lias. This was not realised and indeed for long after the close of the century the Devonian and Carboniferous rocks of Devon and Cornwall were thought to be equivalent in age to rocks below the Old Red Sandstone elsewhere. As for the igneous rocks, the large granite masses were generally believed to be among the most ancient in age, but since their intrusive nature had been demonstrated this assumption was found to rest on no secure basis. However they had not yet been discovered as being intrusive into any but 'primary' rocks. Those smaller intrusions, dykes and sills, found among 'secondary' rocks, were necessarily admitted as being comparatively young. The logical position had thus been reached of a preliminary classification of rocks, regardless of age, into the two groups, igneous and sedimentary.

It is not surprising that no classification of the Primary rocks had been attempted. Particular rocks were simply described by such lithological or mining terms as 'schistus', 'slate', 'killas', and 'clay-slate' which had no very precise meaning. Within the Secondary group a very rough succession from the Coal Measures to the Chalk was, as we have seen (theme 6), given by John Strachey, 1725 [23] and a more accurate one of the same range of strata by John Michell, 1788 [41]. In those regions where the Carboniferous suc-cession had been observed, in Scotland by John Williams, 1789 [43] and particularly in Derbyshire by John Whitehurst, 1788 [37], the

Limestone was found to underlie the Coal Measures, with Millstone Grit (if present) between. It does not seem that the stratigraphical relation of the Old Red Sandstone (the equivalent in age to the marine Devonian) to the Carboniferous rocks had been observed.

The question arises: to what extent could a geological map of Britain have been constructed from the observations recorded up to the year of Playfair's *Illustrations*, 1802 [54]? Something could have been done, but the map would have been very sketchy, and in fact no one made any serious attempt at such a compilation, though William George Maton, 1797 [49] drew a very inaccurate 'mineralogical map' of south-west England (theme *16*).

18 Speculation during the seventeenth and eighteenth centuries

In a general history of geology it would obviously be important to enumerate and discuss the various successive ideas and controversies, in all countries, as to how the earth has come to be as we now know it, both as to its outer features and its inner structure. This is outside our province, which is the growth of sound knowledge about the geology of Britain. It will be sufficient here to cite the following recent works which deal with the cosmologies of the seventeenth and eighteenth centuries: Gillispie (1951, 1959), V. A. Eyles (1955), Chorley and others (1964), Davies (1969), V. A. Eyles (1969).

19 Early manuscript sections and maps

The most important and interesting figure in British geology during the first two decades of the nineteenth century was William Smith, the 'Father of English geology' as Adam Sedgwick called him in 1831. More will be said about him under later headings; we here refer solely to one set of items.

There was an overlap in time between the culmination and full expression of the Huttonian principles and records and the very unobtrusive jottings and sketches of William Smith which were to lead to such great things. Smith's work was an entirely new beginning, a very different kind of research and in the opposite part of the country—Hutton based on Edinburgh, Smith on Bath. Thus it may be said that the years closing the eighteenth century and those opening the nineteenth were the most momentous in the annals of British geology. Incidentally, Charles Lyell was born in the same year, 1797, that James Hutton died.

The earliest of Smith's productions, apart from his notes, are his manuscript sections and maps. Fortunately representative examples of these have been retrieved. Between about 1794 and 1809 [46] he sketched a number of sections of strata showing beds from the Coal Measures near Bristol to those above the Chalk near Norwich. These sections seem to have been the first of their kind. Even more important were his geological maps of the country round Bath, of which the best known is the one dated 1799 [50], the first truly geological map in existence of any part of Britain. At just about the same time Smith made sketch-maps of the whole of England and Wales, the earliest surviving one being that dated 1801 [53]. The viewing of the geology of England and Wales taken as a whole thus begins with the drawing of this map.

William Smith's early sections and maps have been fully set on record and some of them reproduced in the published works of Judd (1897), Sheppard (1917) and Cox, (1942).

Only a very few years later, 1807-8 [59] John Farey drew what are most probably the earliest extended geological sections of British regions; across the north-east Midlands from Derbyshire to the Lincolnshire coast, and across the Weald district of south-east England from London to Brighton. These have recently been brought to light, reproduced in redrawn versions, and discussed by Ford (1967).

20 Palaeontology systematically begun

One work only, but a very important one, stands out conspicuously in the published annals during the first decade of the nineteenth century: James Parkinson's *Organic remains of a former world,* in three volumes, 1804-11 [55]. It is written in the form of letters, and this method of communication to an imaginary correspondent, and thus to the world at large, is surprisingly effective. During the progress of its publication a transformation had been taking place in the human view of the geological scene. This was due chiefly to the dissemination, orally and by correspondence, of the discoveries of William Smith, but also to independent researches, particularly those by John Farey. Parkinson's three volumes reflect this progress. General ideas are grappled with but not always satisfactorily resolved in the first volume, whereas in the third volume correct views, with special reference to fossils and their occurrence, are clearly expounded with full acknowledgement to Smith and Farey. The work is also remarkable for Parkinson's grasp of the previous and contemporary literature of geology and palaeontology and his scholarly discussion of it. Many well-known fossil species are excellently described (with tinted drawings) for the first time, but unfortunately he did not give them binomial (Linnaean) names, so that he does not appear later as the 'author' of these species. Altogether the work has not received the attention its importance deserves.

The term 'organic remains' in the title is of some interest. It was coming to be recognised that the term 'fossil' was being used in the inconveniently wide sense of any noteworthy object 'dug out' of the rocks. 'Organic remains' was appropriate and unambiguous but was found to be rather cumbersome and the term 'fossil' soon afterwards came back into use, thenceforward to refer exclusively to the remains of organisms preserved in the rocks.

In 1809 William Martin [62] published a beautifully printed and engraved quarto volume on the fossils of Derbyshire, all the most noteworthy ones being from the Carboniferous Limestone. About half of this work had already appeared in parts in 1793. In it one of the first attempts is made to apply to fossils the systematic Linnaean nomenclature already being applied to living animals and plants, but Martin uses a variant of the strict form and this has caused some confusion. (Muir-Wood (1951), and Stubblefield (1951), clear up the matter.) Many well-known species, called by the specific names Martin applied to them, such as *Productus giganteus, Spirifer striatus, Lonsdaleia floriformis*—the generic names were applied later—are among his 'figures and descriptions'. In the same year he published another work [63], a small book which embodied so many sound principles in palaeontology that it may perhaps be regarded as the first 'textbook' in English on the subject.

In 1804 James Sowerby [56] began publishing his *British Mineralogy* in five volumes, ending in 1817. This work was in his well-known style (the first volume of his great *English botany* had appeared in 1790). It describes minerals and rock-specimens. His much more important *Mineral conchology of Great Britain* [70] began appearing in numbers in 1812 and went on continuously till 1826, with James de Carle Sowerby taking over in 1822 on the death of his father. The great majority of the many species included were new to science and were given Linnaean names, were beautifully and accurately figured, and not inadequately described. Thus of all citations of authors of the specific names of British fossils the names of these must surely be the commonest. Parts of an unfinished volume appeared in 1846.

J. S. Miller's *Natural History of the Crinoidea*, 1821 [101] is a very notable volume as it is the first British palaeontographical 'monograph' dealing with one well-defined group of fossils.

The first edition of Parkinson's well-known *Introduction to the study of fossil organic remains*, 1822 [107] summarises in detail all palaeontological knowledge so far acquired in Britain and the neighbouring parts of the Continent. It was 'dedicated to the service of those Admirers of Fossils who have not yet entered into a strict examination of the distinctive characters of these interesting substances.'

The works of Parkinson and Martin here referred to have been discussed by the present writer (1948).

21 Experimental geology

Sir James Hall was intimately associated with James Hutton. In the history of general geology he is famous because to him 'we owe the establishment of experimental research as a powerful aid in the investigation and solution of geological problems' (Geikie, 1905(a)). These researches were published in the *Transactions of the Royal Society of Edinburgh* at intervals during the period from 1790 to 1825, the most important being those we have selected, appearing in print in 1805, 1812, and 1815 [58]. Hall's experimental work has been discussed by Geikie and, in detail, by V. A. Eyles (1961). A recent paper on geological experiments is that by Kuenan (1958).

22 The geology of England and Wales: geology in general

Two important books of a general kind were published in 1813. The title of the Rev. Joseph Townsend's very thick volume, *The character of Moses established for veracity* [72], obscures both its contents and its merits. It is chiefly known for having been the means of bringing to light much of William Smith's work but it embodied also the results of his own independent investigations. (Smith's work had already been partly publicised by Parkinson and Farey.) As Woodward has remarked (1911), 'At the time of publication it was the best English work on stratigraphical and topographical geology'. In particuar Townsend should be given the credit for having done original research for, compiled, and published the first work on the stratigraphical palaeontology of Britain; there are 21 plates of fossils characteristic of the successive formations.

Robert Bakewell wrote what from its title, *An introduction to geology* [71], was intended to be, and to some extent actually was, a geological textbook. In it the author makes clear statements on geological structures particularly on the outstandingly important ones of slaty cleavage and unconformity, and moreover uses modern terminology in doing so. A chapter entitled 'An outline of the geology of England' looks promising, but we are disappointed if we anticipate a resumé of the knowledge gained to date. There is a small map showing very roughly and inaccurately three geological regions but it can hardly be called a 'geological map'. Bakewell refers to Townsend's book, just published, but the name of William Smith is conspicuously absent. However he gives good accounts of the regions that he himself had examined; for instance, Charnwood Forest and Titterstone Clee Hill. The book was reviewed by John Farey in the year of its publication.

Greenough's small book, *A critical examination of the first principles of geology,* 1819 [92] comprises essays on such topics as stratification, the inconstancy of lithological character, and fossils and strata. These are topics which at that time (as at any time) required discussion, but, as he admits, he reached little in the way of settled conclusions. It is a legitimate exercise in geological debate and does not deserve to be called 'ill-tempered', as it sometimes is.

In 1818 appeared William Phillips's compilation, *Selection of facts* [90], which was the first effective attempt to gather together in one volume the various geological accounts of the regions of England and Wales that had so far been examined, to give references to those accounts, and to select, co-ordinate, and set out the whole in stratigraphical order. It is thus the first of that long line of indispensable 'geologies' of England and Wales, or Great Britain, without which no general view can be obtained. Phillips had provided a preliminary sketch in 1816 [78]. The 1818 version is specially important for the table of strata contributed by Buckland (see theme *29*).

These two small books were superseded in 1822 by the publication of the famous *Outlines of the geology of England and Wales* by William Daniel Conybeare and William Phillips [103]. Roughly, the first two decades of the nineteenth century may be said to constitute a distinct chapter in the progress of British geology, a chapter which was closed by the appearance of this work. It incorporated and consolidated all previously available information and added much original material. The three regional accounts published in the same year, those of Yorkshire, Sussex, and Anglesey, were none of them referred to. Most of the book was by Conybeare but included many notes supplied by Greenough and the small map is based on the latter's large map (1819) which is the obvious companion work. The Introduction contains an important exposition by Conybeare on the general facts and principles of geology as known up to that time, and a sketch of the progress of knowledge in the science.

This is perhaps a convenient place to make a general reference to the very important series of four papers by North (1931, 1932, 1933, 1934) on the history of geology in South Wales from the time of the writing of Giraldus Cambrensis in 1188 to the time of the founding of the Geological Survey in 1837. One of these papers is devoted to 'Dean Conybeare, geologist.'

23 The geology of Scotland

We have already mentioned some remarks made by Robert Jameson under dykes and sills (theme 8, [52]), the Old Red Sandstone (theme 11, [52]), and unconformity (theme 13, [57]).

In 1807 Louis Albert Necker set out on a three-months expedition through Scotland, making many geological observations which were published as a paper in Geneva in 1809. In 1808 he presented to the Geological Society of London the geological map he had made by colouring a copy of Kitchin's map of Scotland (scale, 12½ miles to the inch) [60]. It must have been for the most part a compilation from the work of others. In 1939 the Edinburgh Geological Society published a colour-printed facsimile of the original, and in 1948 appeared in the *Transactions* of that society the detailed and illuminating paper by V. A. Eyles on the man and his work in Scotland, particularly on his geological map. The map shows very roughly the main features of the modern geological map : the Lower Palaeozoic sedimentary rocks of the Southern Uplands ('Transition rocks, grauwakke, etc.'), the Upper Palaeozoic sedimentary rocks of the Midland Valley and the Moray Firth-Caithness region ('Secondary or floetz rocks, limestone, sandstone, breccia, etc.'), the 'Highlands ('Primitive rocks stratified, as gneiss, mica slate, clay slate'), the granitic intrusions of the Highlands and Southern Uplands ('Primitive rocks unstratified as granite and sienite'), and the volcanic rocks of the Midland Valley and the western island region ('Trap rocks, basalt, wakken, pitchstone, grunstein, etc.'). His names includes many of those in use for the various kinds of rock up to that time, both in Britain and on the Continent.

John Macculloch's *Description of the western islands of Scotland*, in three volumes, 1819 [93] contains an enormous number of significant facts on the geology of western Scotland about which practically nothing was previously known. All the modern papers and memoirs on these regions begin by stating and discussing what Macculloch had to say. Lewisian, Torridonian, Cambrian, Dalradian, Jurassic (with a list of Lias fossils), Tertiary; sedimentary, igneous, and metamorphic; the vitally important structural relationship of unconformity between certain groups; a separate volume of views, sections, and maps. All these (we have of course used modern names) are dealt with at length. As remarked by Geikie, 1897 [318] few single works of descriptive geology have ever done so much to advance our knowledge of the geology of Britain. We quote two passages :

The Cuillin Hills, Skye (one of those spots that are among the highlights of British geology)

'I have remarked in the general description that the Cuchullin mountains are principally formed of a compound to which I have given the name of hypersthene [augite] rock. The difficulty of ascertaining its extent arises from a stormy climate added to the distance from human habitations and the difficulty of access. This rock extends to the mountain boundary of the eastern side of Loch Cornish, as will be seen in the accompanying map. There is even reason to suppose that it forms part of Blaven; an opinion founded on the similarity of its craggy outline and the remarkable nakedness of its rocks.'

The unconformity between the Torridonian and the Lewisian

'It must be remarked with respect to the position of the red sandstone, that it is often unconformable to the gneiss. As it is unnecessary to multiply examples, I need only say, that instances of this relative position occur in Suil Veinn, and in many other parts of the extensive tract under review. In these, the gneiss lies at angles considerably elevated, while the beds of sandstone are nearly horizontal. Wherever the contact of the two is to be observed in these cases, an intimate union will be found to exist; the irregular surface of the gneiss being filled with a breccia formed of its own fragments, strongly adhering as in the instances already mentioned in Lewis and in Rasay, and the stratified structure of the sandstone commencing only after these irregularities are filled.'

Macculloch's geological map of Scotland was published in 1836 (V. A. Eyles, 1937).

In 1820 another important work on the geology of Scotland appeared: Ami Boué's *Essai* [96]. This combines his own original observations which were chiefly in connection with the igneous rocks of Old Red Sandstone and Carboniferous age, with a well-ordered summary of the work of previous writers, particularly Jameson and Macculloch.

All the early geological maps of Scotland, from 1808 to 1858, have been listed by V. A. Eyles (1936).

24 The first series of the Transactions of the Geological Society of London

The year 1811 saw the start of that extremely important medium for the publication of geological researches in Britain: the serial volumes of the Geological Society of London which had been founded in 1807. The first series of the large quarto *Transactions* ran from 1811 to 1821, the second series from 1822 to 1856. After 1845 these were largely superseded by the *Quarterly Journal* which began in that year and has continued to the present day; but from 1871 it is the *Journal* with six issues a year.

Various British regions were described in the five volumes of the first series of *Transactions*. We have selected the following:

Aikin, 1811 [64]. Shropshire. Distribution and characters of the Triassic sandstones, Coal Measures, Silurian limestones, Cambrian quartzite, and the Pre-Cambrian volcanic rocks of the Wrekin—Caer Caradoc—Ragleth ridge

Holland, 1811 [66]. Cheshire. Many important geological facts recorded for the first time: the succession of strata tabulated; the salt in interbedded deposits not as veins; the very important negative fact of the absence of fossils is noted. Holland shows however that he has little conception of the length of geological time.

Horner, 1811 [67]. Malvern Hills. Pioneer work. (See theme *46*)

Parkinson, 1811 [68]. London. The first detailed account of the strata of the London basin. It is particularly important as being the first application, in detail and in a particular region, of the principle that strata are characterised by their own peculiar fossils. Parkinson refers particularly to William Smith whose researches were already known by hearsay in England, and to Cuvier and Brongniart in France.

Webster, 1814 [75]. Isle of Wight. (See theme 26)

Berger, 1814 [74], and Henslow, 1821 [100]. Isle of Man. Berger gives a pioneer account, with emphasis on mining, with a geological map and sections. Henslow improves on this with a greater amount of purely geological detail. Lamplugh, 1903 [350] gives appreciations of both.

Horner, 1816 [77]. South-western Somerset. This is accompanied by a very good geological map, more accurate over this area than either Smith's or Greenough's.

Bright, 1817 [81]. Bristol. The Carboniferous Limestone section in the Avon gorge and the unconformable relationship between the Carboniferous and overlying Mesozoic

Buckland, 1817 [82]. Appleby. A description of the Cross Fell inlier of Lower Palaeozoic rocks and its geological setting

Buckland, 1817 [83]. South-east England. A particular account of the Lower London Tertiaries ('Reading beds', 'Plastic clay', etc) of the London and Hampshire basins

Macculloch, 1817 [85]. Glen Roy. The first close examination and reasoned discussion of the famous 'parallel roads'. Macculloch recognised that the beaches were those of a lake which suddenly subsided three times, due to some local cause.

Winch, 1817 [88]. Northumberland and Durham. A number of interesting observations : Magnesian Limestone unconformable on Coal Measures; fossils from the Magnesian Limestone; the non-marine character of Coal Measures fossils; an igneous dyke and its metamorphism of coal seams; and an account of the Whin Sill—its mineralogical character, geographical occurrence, and its effect in producing waterfalls.

Hennah, 1817–25 [84]. Plymouth. Fossils in the Devonian lime-stone particularised, with localities.

Phillips, 1819 [94]. Dover. The stratal succession in the neigh-bouring cliffs, the Lower Cretaceous sequence in particular begin-ning to be realised.

Buckland, 1821 [98]. Lickey Hill, Worcestershire. A clear des-cription of a neatly-significant geological structure, a little inlier of Pre-Cambrian (?) and Lower Palaeozoic rocks surrounded by an unconformable cover of New Red Sandstone.

25 *Derbyshire geology and the northern Pennines: stratification, structure, section, map, column*

Farey's *Agriculture and minerals of Derbyshire,* 1811 [65] is one of the classics of British geology. The title does not promise much in the way of general geological interest which is perhaps why it has been largely overlooked. The main contributions to geological knowledge contained in the volume may be summarised as follows :

A fundamental principle of structural geology, that of faulting with subsequent denudation, is established and worked out with great thoroughness and illustrated with coloured block-diagrams, and the factors governing outcrop are clearly explained.

The succession of the Carboniferous rocks in Derbyshire, particularly the middle portion, is described in detail.

In describing the rocks of Charnwood Forest in Leicestershire, Farey gives a very early account, perhaps the first, of a British region now known to be largely composed of Pre-Cambrian rocks.

Probably the most important part of the whole book is his 'Map of Derbyshire and parts of the seven adjacent counties showing the principal strata and soils' (scale, 6¼ miles to the inch). It is one of the earliest of the truly geological maps of any part of Britain to be published and appears to be entirely original. It is only necessary to compare this map with the modern quarter-inch map of the Geological Survey to realise its quality, though he was incorrect in his structural interpretation of certain of his lines and in his stratigraphical assignment of the Trias.

The book has been extensively considered by the present writer (1947). The following are a few quotations :

'The strata [Coal Measures] all contain seams of coal, separated by numerous strata of bind, clunch, shale and other argillaceous strata which enclose the coaly impressions of vegetables.'

'Around Ashover one of the most perfect specimens of local denudation [of a dome-shaped fold-structure] is to be seen that can perhaps anywhere be witnessed.'

'The Millstone Grit rock by its thickness and its hardness gives rise to the greater part of the siliceous rock scenery in Derbyshire and the adjacent parts of Staffordshire, Cheshire and Yorkshire.'

In 1811 [69] and 1813 [73] White Watson published two interesting little books describing Derbyshire strata, with a section across them. The section is nearly two feet long; the general broad anticlinal structure is well shown, but the vertical scale is so enormously exaggerated as to distort the natural arrangement. Ford (1960) has recently thrown a flood of light on Watson and his work.

The convenient method of showing graphically a stratigraphical succession by a geological column was already in use at the end

of the eighteenth century (see for instance the frontispiece to
Arkell and Tomkeieff's *English rock terms,* 1953). Westgarth
Forster's book *A treatise on a section of the strata, from Newcastle-
upon-Tyne to the mountain of Cross Fell,* 1809 [61] is notable for
a very long column, some 5½ feet long altogether, which is shown
in portions and comprising a total thickness of strata of '1345 yd.
oft., 10in.' It is a typical column or 'vertical section', in the Geological Survey's meaning of the term, showing by lines and dots the
various kinds of rock, and drawn to scale. On one side are written
the 'local names' (which include such curiosities as 'girdles', 'post',
'thill', 'hazle', 'plate', 'tuft') and on the other the 'nature of each
stratum' (more ordinary adjectives, such as 'argillaceous', 'siliceous',
'bituminous').

26 The Isles of Wight and Purbeck

Thomas Webster's long and detailed paper, 1814 [75] on the
Tertiary formations of the Isle of Wight set a new standard for the
Geological Society's *Transactions.* In it he described in particular
the Oligocene strata, showing them to comprise two freshwater
formations alternating with marine. He was applying a principle
that had recently been apparent in France, that the marine or freshwater conditions of deposition of beds of this age could be inferred
from the kinds of shells in them. We have already seen in theme *4*
that this kind of inference had been drawn with regard to the Coal
Measures. This account together with that of the general geology
of this island and the Isle of Purbeck (1816) constitute a great
stride forward in the quality of British descriptive geology. Webster,
in his contribution to Englefield's handsome volume, 1816 [80]
gives an account of a 'classical' region of Cretaceous and Tertiary
formations and of a striking regional geological structure hitherto
unknown. In spite of a few understandable misreadings it remains
unsurpassed as a lucid, forceful, and entertaining account of Wight
and Purbeck. Webster was an artist and most of the splendid plates
in the book are from his drawings, nearly all of which are of
geological features. There are also coloured panoramic views of the
coast showing the arrangement of the strata. Most important of all
is the coloured geological map of the stretch of country from the
Isle of Wight to the Isle of Portland (scale, 2 miles to the inch).
The appearance of the map is a landmark in the history of the
geological mapping of Britain. Webster's writing is in the form
of letters to Sir Henry Englefield and there are twelve of these,

written at various times during the years 1811, 1812, and 1813. These letters have been reviewed by the present writer (1949), and we here transcribe the first of them :

'On arriving at Parkhurst I proceeded to St. Helens. I hired a boat next morning with the intention of sailing round Culver cliff and landing in Sandown bay. The cliffs at Bembridge consist of clay and gravel [the raised beach] resting on a bed of calcareous stone [Bembridge Limestone] which forms a ledge of rocks extending a considerable way eastward. Having passed through these shoals, I landed in Whitecliff bay, where the strata are quite vertical, consisting of alternations of clay and sand [Eocene, with the Headon and Osborne Beds of the Oligocene], the bed next the chalk being a deep red colour [Reading beds]. I followed these cliffs northwards until I came to the calcareous bed, which I had already seen at Bembridge ledge. Here it is nearly horizontal, but when it approaches the vertical clay strata it rises towards them and when very near turns upwards, forming a curve. The chalk strata next to the clay are nearly vertical. Taking now to my boat to see the true dip of the strata in the face of the cliff, I observed that the lines of flints on the north side dip to the north at about 70°, and that the dip lessens towards the south side, where it is about 50°. Doubling this cape, the view suddenly opened into Sandown bay. Proceeding westward, I observed that the flints disappeared and that the inferior bed of chalk was wholly without them. From below this, other strata emerged, the whole thing being highly inclined and exhibiting in the most instructive manner a regular succession of the beds.

Immediately below the chalk without flints is a stratum of yellowish white marl or argillaceous chalk [Grey Chalk and Chalk Marl]. A thick stratum of sandstone succeeds, consisting of siliceous sand and united by calcareous matter and containing green earth [Upper Greensand]. The next stratum consists of a dark blue or grey marl [Gault], its place being marked by a considerable hollow. This is followed by a succession of dark red and yellow ferruginous sands [Lower Greensand] forming the Red cliff, which bears a striking contrast to the cliffs in the neighbouring chalk. Near the termination of these cliffs are several thin strata [Wealden] of a stone composed of bivalve shells in a calcareous matrix, much resembling Portland stone.'

Webster's struggles to vindicate and enlarge his researches in southern England are apparent in his correspondence (Challinor,

1961-64). His work in the Isle of Wight has also been reviewed by Jackson (1932).

Forbes's Geological Survey memoir, 1865 [174] *On the Tertiary fluvio-marine formation of the Isle of Wight* (the Oligocene as it was soon to be generally called; the name had been proposed two years before), carried on the survey, with an analysis of Webster's accounts and of the papers that had been written since Webster's time in the island. This was followed by the general memoir and accompanying map by Bristow in 1862 [188]. A second edition, revised and enlarged by Clement Reid and Aubrey Stratham was issued in 1889. Beautiful sets of 'horizontal' sections (scale, 6 inches to the mile) correspond to each of these editions. Meanwhile there had been Judd's paper on the 'Oligocene' strata (1880). In 1921 was issued the *Short account,* by H. J. Osborne White, based on the 1889 memoir; and both the Isle of Wight and the Isle of Purbeck are covered in the *Regional geology* handbook on the *Hampshire basin* 3rd edition, 1960 [548].

Two important works on the Isle of Purbeck are in our selection, both detailed memoirs of the Geological Survey. Strahan's 1898 [325] comes just within the nineteenth century, while Arkell's, 1947 [586] (see theme *71*) is fifty years later.

27 The first published geological maps and sections of England and Wales

We have already seen that William Smith's discoveries were becoming generally known among the savants, and many of them had been published in the works of Parkinson, Farey, Townsend, and the Rev. Richard Warner.

It was not until 1815, and then after much delay, that anything geologically significant was published by William Smith himself. This was his great map, the *Delineation of the strata of England and Wales* [76], on the large scale of 5 miles to the inch. It was issued in some cases as an atlas and in others in rolled form. It is the first geological map of England and Wales, the basis of all subsequent geological maps of the country, wonderfully detailed and accurate, especially as an achievement of one man. His *New geological map,* 1820 [97] on the scale of 15 miles to the inch incorporates a few slight additions to the information on the large map. Between 1819 and 1824 were issued six parts of his *New geological atlas of England and Wales* [95]. Each part contained four sheets, one to a county, except Yorkshire, which is in four sheets. The scale varied from 2 to 4 miles to the inch. This atlas of large-scale maps was never completed.

In 1819 a large geological map of England and Wales was published under the auspices of the Geological Society [91]. George Bellas Greenough was chiefly responsible for its production but several of the leading geologists of the time, particularly Buckland, had a hand in it. It has always been known as 'Greenough's map'. There was nearly as much delay here as there was with Smith's map. Both were on much the same scale. Smith got in first but understandably the 1819 map was the better of the two. It is not clear to what extent it was based on Smith's map, but that it largely was so was eventually acknowledged in the third and last edition of 1865. Nevertheless it was fuller and more accurate than Smith's even in the region of the counties of Oxford, Gloucester, and Somerset, with which Smith was most familiar.

Smith's work, and his maps in particular, with the circumstances of their production, has been much commented on. See especially Fitton (1818), Phillips (1844), Judd (1898), Sheppard (1917), V. A. Eyles and Joan M. Eyles (1936), Cox (1942), Cox (1948), Davis (1952), Joan M. Eyles (1969 a and b). For commentary on Greenough's map : Conybeare (1832), Judd (1898), Woodward (1908), North (1928).

Nine charts of sections by Smith across various parts of England and Wales were issued during the years from 1817 to 1824 [87]. These were (1) London to Snowdon, 1817, 'one of the most remarkable of Smith's productions . . . measures $53\frac{1}{4}$ inches by 12 inches' (Sheppard, 1917); (2) south of London, 1817; (3) North Wiltshire, 1817; (4) East Anglia, 1819; (5) through Hampshire and Wiltshire to Bath, 1819; (6) Essex, Hertfordshire, and Cambridge, 1819; (7) the Weald, 1819; (8) Dorset and Somerset, 1819; (9) Yorkshire, 1824. These have been fully described by Sheppard (1917), Cox (1942), and Joan M. Eyles (1969 a), and nearly all of them reproduced by either Sheppard or Cox.

28 Strata characterised by their fossils

In a note jotted down at the Swan Inn, Dunkerton, Somerset, on 5 January 1796, William Smith made what seems to be his first written statement about 'that wonderful order and regularity with which Nature has disposed of these singular productions [fossils] and assigned to each class its peculiar stratum' (quoted by Phillips, 1844). He makes a similar statement in the *Memoir to the map*,

1815 [76] and in the same year, in making observations on his map to the Society for the Encouragement of Arts, he refers to fossils as 'indices' : '. . . the present discovery of regularity in the strata, to which fossil shells are now become the Indices' (quoted by Davis, 1943). Ever since Parkinson publicly drew attention to the fact in 1811 in both the *Organic remains* [55] and the *Transactions* paper [68] Smith has been deservedly accorded the honour of having first thoroughly grasped this 'law' about fossils and strata, stated it, applied it, and illustrated it. His most important publication, another of the classics of British geology, was his *Strata identified by organized fossils* [79], which was issued in four parts (of a projected seven) during the years 1816 to 1819. The title continues : 'containing prints on colored paper of the most characteristic specimens in each stratum.' His other work was his *Stratigraphical system of organized fossils,* 1817 [86] of which the title continues : 'with reference to the specimens of the original geological collection in the British Museum; explaining their state of preservation and their use in identifying the British strata.' This collection, of nearly 3,000 fossils (of which the work is a catalogue), was bought by the British Museum in 1816 and is still preserved at South Kensington (Joan M. Eyles, 1967).

29 Early tables of strata

In 1799 William Smith dictated to his Reverend friends at Bath, Benjamin Richardson and Joseph Townsend, his now famous list of the English strata [51]. This was published by Fitton (1833) and Phillips (1844), and Sheppard reproduced the original manuscript in facsimile (1917). An earlier list (1797) has been discovered (Douglas and Cox, 1949), very similar to the 1799 list. These two lists are the earliest reasonably comprehensive and detailed tables of the British strata.

Smith's stratigraphical lists were published in several forms by himself and by others. Of our selected works they occur in Farey, 1811 [65] and Townsend, 1813 [72]. They are also enumerated in Richard Warner's *History of Bath* (1801) and his *New guide through Bath* (1811). Smith himself gives a list with his map, 1815 [76] and his catalogue, 1817 [86]. These lists are substantially similar but show some additions and emendations as time goes on (the 1797 and 1799 lists do not go higher than the Chalk), with the gradual introduction of the well-known stratal names. The lists are much more detailed for the lower part of the Jurassic than for

G

the rest of the succession, as that was the part that Smith knew best.

William Buckland [89] contributed to William Phillips's *Selection of facts* (1818) an important table of the British strata. This tidies up the Smithian lists in some details, is fuller in the Tertiary section which was based on Webster's work and adds the 'Old Red Sandstone' below the (Carboniferous) 'Mountain Limestone'. A very similar version of this table is given by Conybeare and Phillips, 1822 [103]. The list accompanying Greenough's map, 1819 [91] is also important.

In all these lists there is a very imperfect realisation of the succession of the Cretaceous formations below the Chalk, and all the Lower Palaeozoic is indicated by such names as 'killas', 'slate', 'limestone'. (For a comparative table see Challinor, 1970.)

30 The Lake District

Jonathan Otley, in his account of the geology of the English Lake District in 1823 [110], outlined with a sure hand and with practically no previous knowledge to guide him, all the salient features of one of the most geologically interesting regions of Britain. His account is very simply but very carefully written, and a completely unaffected literary charm is added to its purely scientific value. His account may perhaps be said to form a trilogy with Farey's of Derbyshire, 1811 [65] and Webster's of the Isles of Wight and Purbeck, 1816 [80]. Each of the three was a pioneer but detailed description, and a largely accurate interpretation, of geological features. The researches were quite independent but they were nicely separated geographically and geologically. Between them the rocks described ranged from the Pre-Cambrian to the Recent, with typical representatives of nearly all the stratigraphical systems, and all the major structural phenomena in geology were recorded by one or other of the three. Here, in Otley's little book, is the first account of a British Lower Palaeozoic region. He established the succession of what we now know as the (Ordovician) Skiddaw Slates, Borrowdale volcanic series and Coniston Limestone, and the succeeding Silurian.

'The rocks of which these mountains are formed may be classed in three principal divisions. Of these, the first or lowest in the series forms Skiddaw, Saddleback [etc]. The second division comprehends the mountains of Eskdale, Wasdale,

Ennerdale, Borrowdale, Langdale [etc.], including Scawfell and Helvellyn. All our fine towering crags belong to it. The third division, forming only inferior elevations, commences with a bed of dark-blue or blackish transition limestone, containing here and there a few shells and madrepores. It dips to the south-east where succeeds a series of rocks principally of a slaty structure.'

He gives particularly correct, clear, and elegantly expressed statements relating to unconformity, and the distinction between cleavage and stratification. 'The superincumbent bed of limestone mantles round these mountains, in a position unconformable to the slaty and other rocks upon which it reposes'. This refers to the dominant structural feature of the region, the unconformity (ring-shaped in outcrop) between the Carboniferous (or Permo-Triassic) rocks and the Lower Palaeozoic.

'The most natural position of the lamina or cleavage of the slate appears to be vertical, but it is to be found in various degrees of inclination, both with respect to the horizon, and the planes of stratification. The direction of slaty cleavage bears most commonly towards the north-east and south-west; while the dip or inclination varies.'

His other observations range from igneous plutons and dykes to erratic boulders. Otley's account of the geology of the Lake District has been the subject of a detailed analysis and commentary by the present writer (1951).

Adam Sedgwick carried out most of his work in the Lake District during the period 1822 to 1824, but the results were not given to the world till much later, in papers to the Geological Society's *Transactions*, 1835 [132], *Quarterly Journal*, 1845-8 [149], and in his letters addressed to the poet Wordsworth, 1842-53 [153]. In these papers and elsewhere Sedgwick paid tribute to Otley. He himself greatly amplified Otley's work, subdividing and naming the subdivisions of the rock-groups, particularly those of Otley's 'third division', and suggesting correlation with other areas (detailed in Marr's *Geology of the Lake District*, 1916 [416]).

31 'Diluvian remains': Pleistocene mammals

In 1823 a work was published, very different from Otley's of the same year. This was William Buckland's well-known *Reliquiae diluvianae* [109]. As in the case of Townsend's book ten years earlier (both authors were geologist-divines), a biblical title obscured the value of the work from a purely geological point of view. Leaving aside Buckland's thesis that he was endeavouring to support by geological evidence (for this see, particularly, Gillispie, 1951), we here record that he brought to light, with a thoroughly scientific industry and perception, the existence of a great number of bones of extinct mamals from caves and river-gravels, particularly from the Kirkdale cave in Yorkshire.

> 'The bottom of the cave, on first removing the mud, was found to be strewed all over like a dog-kennel, from one end to the other, with hundreds of teeth and bones, or rather broken and splintered fragments of bones, of all the animals above enumerated [23 kinds, including the hyaena, tiger, bear, elephant, horse, ox, deer, rabbit, mouse and several birds]. It must appear probable that the cave at Kirkdale was, during a long succession of years, inhabited as a den of hyaenas, and that they dragged into its recessess the other animal bodies whose remains are found mixed indiscriminately with their own.'

32 Discovery of Mesozoic reptiles

It is not very easy to trace the documentation of the early discovery and naming of those great groups of extinct reptiles which dominated the life of the Mesozoic. We here give some details.

The most important work being done in all branches of vertebrate palaeontology during the earlier part of the nineteenth century was that by the great French naturalist, Baron Georges Cuvier. The British discoveries were thus largely to be studied in the light of Cuvier's slightly earlier, more comprehensive, and more detailed researches. From 1796 onwards his papers appeared in the *Annales du Muséum* and were collected in the four-volume *Recherches sur les ossemens fossiles,* 1812. This work passed through four editions and for our purpose, at least, the most important of these is the 'nouvelle edition' of 1821-4. This is in five volumes, the last of which is in two parts, and it is the second part of volume 5 (1824) that deals with the reptiles. It contains many references to the British activities.

The discoveries of fossil reptiles made before 1811 were reviewed by Parkinson in the third volume of his *Organic remains of a former world,* 1811 [55] largely on the authority of Cuvier. We have already mentioned Stukeley's ichthyosaurian, 1719 [24] and Chapman and Wooller's crocodile, 1758 [32] (theme *3*), but the most notable fossil brought to light before 1811 was 'the large animal' from the Chalk near Maastricht in Holland. This was described by P. Camper in the *Philosophical Transactions* in 1786. Its name, *Mosasaurus,* was proposed by Conybeare in Cuvier's 1824 volume, and it has been placed in an extinct group among the Squamata (lizards and snakes) by Zittel (1932). Parkinson mentions reptilian remains that had been found in the cliffs of the Dorset and Yorkshire coasts, and near Bath.

This year, 1811, saw the beginning of the era of systematic collection and scientific examination of British vertebrate fossils. Important finds were made by Mary Anning of Lyme Regis, the romantic circumstances of which have been given by De la Beche (1848), Elizabeth Gordon (1894), and Woodward (1908).

Fragmentary bones of what are now to be recognised as those of the aquatic 'saurians' *Ichthyosaurus* and *Plesiosaurus* had been figured in various early works, particularly by Lhwyd, 1699 [17]. The newly-discovered specimens of *Ichthyosaurus* were first described by Sir Everard Home (1816-20), with William Clift's 'accurate and beautiful engravings' (Owen, *Mesozoic fossil reptiles,* 1881 [163]). The name *Ichthyosaurus* was given by König of the British Museum in 1818, and its adoption was advocated by Conybeare, but Home had used the name *Proteosaurus. Ichthyosaurus* and *Plesiosaurus* were fully described by Conybeare, 1821 [99]; 1822 [102]; 1824 [113], the name *Plesiosaurus* being given in his 1822 paper.

The land 'saurians' known as 'dinosaurs' were first found in Britain. This priority is stated by Zittel (1901) and by Swinton (1934). Famous discoveries were those of *Megalosaurus* by Buckland, 1824 [111] and *Iguanodon* by Mantell, 1825 [114] actually first found by Mrs. Mantell in 1822. There is more about Buckland and the *Megalosaurus* by Buckland himself, 1836 [133] and by Mrs. Gordon (1894), and more about Mantell and the *Iguanodon* by Mantell himself (1833) and in his journal edited by Curwen (1940), and by Colbert (1962).

Flying reptiles, Pterodactyls or Pterosaurs, were first found abroad, particularly in the Lithographic Limestone (Ur Jurassic) of Bavaria. The generic name *Pterodactylus* was given by Cuvier in 1809. Buckland, 1829 [121] described a new species found by Mary Anning from the Lias of Lyme Regis.

We may mention here that the first discovery of mammalian remains (now unquestionably known to be such) from rocks older than the Tertiary was made by Buckland, and they were, most surprisingly, from rocks as old as the Stonesfield Slate, far back in Jurassic time. The discovery was announced rather tentatively in his 1824 paper on *Megalosaurus,* and some of the circumstances of identification have been detailed by Owen (1846).

A review of what was known about fossil reptiles up to 1830 was given by Edward Pidgeon. The descriptions were based on Cuvier's but many of the historical allusions were apparently his own.

Excellent accounts of the different kinds of fossil reptiles are to be found in the two handbooks of the British Museum (Swinton, 1965 and 1967).

33 *Coastal geology*

Four of De la Beche's five papers in the Geological Society *Transactions,* 1822-9 deal with the stretch of Dorset-Devon coast from Bridport to Torquay, the fifth, 1826 with South Pembrokeshire. Of the south-coast papers the most notable is the first, 1822 [104] chiefly because of its extended panoramic view of the geological section exposed all the way from Bridport to Babbacombe Bay. It shows the conformable downward sequence of Inferior Oolite, Lias, and Trias, with their slight easterly dip, gradually following one another along the coast westwards, with the almost horizontal (Cretaceous) Upper Greensand and conformably overlying Chalk in the higher places forming the upper parts of the cliffs. The (unconformable) overstep of the Cretaceous, from its position on the Inferior Oolite at the east end of the section to its position on the Trias at the west end is very well shown, but De la Beche does not seem to realise this structural situation clearly. However the section is reproduced, re-drawn and on a smaller scale in his *Sections and views* (1830), where the situation (which he calls 'overlap') is realised. De la Beche mentions the fossils from the Lias—reptiles, fish, ammonites and other molluscs—the most important record, with description and full-page drawing, being that of the fish *Dapedium,* here discovered for the first time.

Two papers of 1826 describe, one the Lias [116], the other the Greensand and Chalk [117], at Lyme Regis, a feature of these being the detailed geological columns shown in the plates. By this time many of the fossils had been described by Sowerby, and De la Beche refers to his figures. A similar paper, 1829 [122] describes, with small coast sections and maps, the Torquay area of Devonian and New Red Sandstone. As the formations of this area were practically unknown and are complicated in their structure, it is not surprising that this paper is unsatisfactory. In particular he misread the Devonian limestone as Carboniferous. It remained for De la Beche himself to clear up the matter ten years later.

The third 1826 paper [115] (the first in that volume of *Transactions*) deals with South Pembrokeshire. There is a good map showing the outcrops of the 'greywacke', 'Old Red Sandstone', 'Mountain Limestone', and 'Coal Measures', and a series of coast-sections, one of which shows the contortions which are such a feature of the Upper Carboniferous cliffs between Tenby and Saundersfoot, with a sketch of Monkstone Point in the text.

The paper in the *Transactions* for 1835 [130] on the Weymouth area, by De la Beche and Buckland, is fuller than the others. There is a large-scale map (1 inch to the mile) of the country between Ringstead Bay on the east and Lyme Regis on the west, with sketch-sections and columns.

The above six papers dealing with south-western Britain may be considered as constituting one whole. With this set the deservedly more famous book on the coast and eastern part of Yorkshire, away to the north-east of the country, by John Phillips, 1829 [123] forms a pair. His work is much fuller and indeed sets a new standard in regional geologies. There is a geological map, sections, and columns in the style of De la Beche's papers, but here they are more detailed. There are fourteen plates of fossils, many of which were new to science and to these he gives names, but he does not describe them. The second edition of 1835 is very little altered.

The panoramic views of the geology of the coasts form a most informative and attractive feature of all the works considered under this heading. (We have seen, in theme 26, that Webster drew the same thing for the Isle of Wight in 1816). A modern atlas of such view-sections for the coasts of the whole of Britain might be a very useful and interesting production.

34 Regional accounts

Three noteworthy regional accounts all appeared in the same year, 1822, one of the most productive years in the annals of British geology. John Stephens Henslow's paper on Anglesey [105] in the first volume of the *Transactions of the Cambridge Philosophical Society* is similar in style and format to the more elaborate of those appearing at about this time in the Geological Society's *Transactions*.

The most noteworthy feature of it is the fine double-page geological map on the scale of about 2½ miles to the inch. As Greenly has remarked in the Anglesey memoir, 1919 [424], 'It shows a high degree of scientific penetration, as well as painstaking research, and on his map his lines have already assumed the same general aspect that they have today.'

Gideon Algernon Mantell's *Fossils [and] geology of Sussex* [106] is a large quarto volume with 42 plates, mostly of fossils, with the description of which the work is chiefly concerned. He is the author of the well-known generic names *Ventriculites* (a sponge) and *Marsupites* (a crinoid). He describes the stratigraphy of the whole of the Cretaceous succession and up to and including the London Clay, clearly recognising the Lower Greensand (so-called by Webster two years later) as a separate formation. He has a map which is an advance on the corresponding part of Greenough's map (1819). Much of this work was incorporated in his *South-east England* (1833).

George Young's *Geological survey of the Yorkshire coast* [108] is also a substantial volume, with tinted engravings, chiefly of fossils 'by John Bird, artist'. While giving a good deal of local information it adds little of general significance to what was already known.

Apart from the works already considered we have selected the following as being among the more important regional descriptions published during the eighteen-twenties and eighteen-thirties.

Buckland and Conybeare, 1824 [112]. Forest of Dean and Bristol coalfields.

A very detailed account, particularly of the Bristol neighbourhood, with sections, and a map (scale 2 miles to the inch). Includes an enumeration of the Carboniferous Limestone beds in the Avon gorge (with the assistance of De la Beche) and a sketch of the striking and well-known unconformity of Jurassic Oolite on Carboniferous Limestone in Vallis Vale. We also find what is apparently the first usage of the term 'anticline', in the form 'anticlinal lines', 'formed by the saddles of the strata'. Immediately following this paper is one by Thomas Weaver with its account of the 'Transition tract' (Lower Palaeozoic) of the Tortworth inlier.

Murchison, 1827 [118], 1828 [120]. Brora coalfield in Sutherlandshire and other stratified rocks, chiefly Jurassic, in the north of Scotland.

At the instigation of Buckland and Lyell, Murchison described this 'coalfield', with its one seam of coal three or four feet thick, and showed as they had thought, that the coal and associated strata were of the age of the Jurassic Lower Oolites.

Martin, 1828 [119]. Part of Sussex

'Filled up the gap left between the eastern part of Sussex explored by Dr. Mantell, and the extreme west of Sussex surveyed by Sir R. Murchison' (Topley, Weald memoir, 1875). This paper deals with Cretaceous strata below the Chalk. It was read to the Geological Society in 1827 and published separately a year later.

Sedgwick, 1829 [124], 1836 [136]. Permo-Trias (Magnesian Lime-stone and New Red Sandstone) of the north of England

Sedgwick did the field work in the years 1821 to 1823 but the results were not worked out for five years, his papers being read to the Geological Society in 1826, 1827, and 1828 (these three published in 1829) and a fourth in 1832 (published in 1836). In 1887 in Woodward's *England and Wales* (2nd edition) it could be said 'We owe our first and best descriptions of the Permian rocks to Prof. Sedgwick'.

Smith, 1832 [128]. Hackness Hills, Yorkshire

The last important work by William Smith. A remarkably detailed and accurate lithographed map on the scale of about 6 inches to the mile. The strata shown are of the middle part of the Jurassic. It is described and reproduced by Sheppard (1917). The memoir by Fox-Strangways was first published in 1892 [294].

Maclaren, 1839 [140]. Fife and the Lothians

One of the most perceptive of the early works on the geology of Scotland, particularly in its recognition and description of the volcanic rocks of the Pentland Hills (Old Red Sandstone) and of Edinburgh itself (Carboniferous).

Prestwich, 1840 [143]. Coalbrookdale coalfield

When Murchison learnt that Prestwich was working on the geology of this area, which lay towards the north-east corner of his 'Silurian' region, he left the matter to him, so that Prestwich's very full paper in the Geological Society's *Transactions* may be taken as a piece of collaboration. Many details of the stratigraphy and structure are recorded, but the Symon 'fault', a locally strong un-conformity within the Coal Measures for which the coalfield is famous, is not shown on his sections. The fossil plants and, particularly, invertebrates, are listed; they are drawn by J. de C. Sowerby and mostly described by him. The most notable of these is the xiphosurian arachnid, named *Prestwichia* some thirty years later and familiar by that name, but recently changed unfortunately to *Eupröops* owing to a rigorous application of the laws of zoological nomenclature which of course apply also to fossils.

Ramsay, 1841 [147]. Isle of Arran

This slim volume 'has long since taken its place among the classics of Scottish geology' (Geikie, 1895).

35 Treatises on geology: principles, elements, and constitution of the science

Lyell's famous *Principles of geology*, 3 vols, 1830-3 [125] produced a greater influence on the progress of the science than any
work before or since. It went through twelve editions altogether
which can be grouped into three sets (Challinor, 1968) : 1-5, 1830-7;
6-9, 1840-53; 10-12, 1867-75. It was in fact a very comprehensive
and elaborate treatise, gathering materials from many countries.
Four kinds of subject-matter are incorporated. The first is contained
in the preliminary chapters which give an historical review, particularly of theories and ideas. The second is the general philosophical principle of Uniformitarianism (Lyell does not use this
word) which lies at the back of the whole work and is incorporated
in the subtitle which, in the first five editions can be paraphrased
as 'the present is the key to the past', and in the remaining editions
as 'the past is the key to the present'. His statements on this are
scattered and somewhat obscure and do not add very much to
Hutton's pronouncements, but he brought to the fore the general
idea of the continuity of the causes of geological change, past,
present, and future. The third kind of subject matter occupies most
of the work. It is a very thorough examination of the working of
the many kinds of geological process, exemplified from all over the
world. Although the author had in mind all the time that all this
was being used to illustrate his principle of uniformity, it is the
expositions of the main facts of geological nature that are so important. The fourth component of the whole work is the description
of the Tertiary strata and their fossils, with some notice of the
Secondary (Mesozoic) formations. This ends with a table of strata
which, down to the base of the Old Red Sandstone, is substantially
that of today. After the first five editions had appeared Lyell
separated much of the factual material, including all our fourth
component (much expanded downward through the Mesozoic into
the newly-investigated Lower Palaeozoic), but excluding most of
physical geology, to form a separate work, the *Elements of geology*,
1838 [138]. This ran through six editions, the last (1865) being
greatly enlarged.

In 1831 De la Beche's *Geological manual* [126] was published,
with a third edition in 1833. It was thus contemporary with Lyell's
Principles. Lyell refers to De la Beche's book here and there in his
second and third volumes and De la Beche similarly refers to the

earlier part of Lyell in his third edition, but the two works are essentially quite independent. Each covered much the same ground, the whole domain of geology as then known. De la Beche's book is better organised that Lyell's and his ideas on some matters were the sounder, particularly in his realisation of the potency of subaerial over marine erosion in carving out hills and valleys. It also gives a fuller account of the facts concerning the successive stratigraphical formations, both at home and abroad, than is given in the first edition of the *Principles*. Thus Lyell's work with its stimulating and debatable philosophy, just right for the time, and its compelling form of presentation in description and picture, at once became famous, overshadowing De la Beche's more formal treatment. (Incidentally, but perhaps not wholly irrelevantly, the printing and binding of the Lyell volumes were more attractive than those of De la Beche's volume.)

In 1834 there appeared the first of the series of textbooks by John Phillips. This was the *Guide to geology* [129], a small volume, but so orderly and comprehensive as to be noteworthy. (Phillips had expounded some general geology in the first edition of his *Yorkshire coast*, 1829 [123] which was omitted in the second edition of that work, 1835, as being by that time 'unnecessary'.) The second edition of the *Guide* (1835) contains 'a new plate of a peculiar kind'. This is a perspective drawing of a geological model of the Isle of Wight showing the surface outcrops and cut through to show the structure in section. We have now had exemplified, in our progress through the years, all the usual forms of graphical geological representation. This work was followed by the much fuller (two small volumes, but 642 pages in all) *Treatise on geology*, 1837-9 [137].

There have been innumerable discussions of Lyell's work. First to be mentioned are his own letters (edited by Katherine M. Lyell, 1890). We may refer particularly to the following: Fitton (1839), Bonney (1895), Judd (1911), Gillispie (1951), Bailey (1962), Chorley and others (1964), North (1965), Davies (1969), and of course the general histories of geology.

During these years of the eighteen-thirties we have the essentials of modern geology firmly established. The works above referred to consolidated all that had been known before, introduced a vast array of new facts, put forward general principles in geological philosophy, and thus constituted a firm basis for the future. The position was somewhat similar to that of some forty years before. We then had Hutton and Playfair establishing the advance, while William Smith came from behind and pushed forward along a quite independent line. And now we have two geologists who, taken

together, were about to do something of the same kind : Adam
Sedgwick and Roderick Murchison.

36 Cambrian and Silurian

It was in 1831 that Sedgwick and Murchison began their ex-
plorations among the Lower Palaeozoic rocks of Wales and the
neighbouring counties. The time was ripe and the region was obviously inviting. Murchison took the Welsh Borderland and worked
downwards from the base of the Old Red Sandstone through a comparatively orderly succession of strata. Sedgwick meanwhile
grappled with the more intractable mountains of North Wales, working upwards from the lowest rocks which, with remarkable insight,
he had recognised as occurring in northern Caernarvonshire.
Henslow's *Geology of Anglesea,* 1822 [120] probably pointed the
way.

It may here be remarked that Sedgwick included in his Cambrian
certain rocks in north-west Caernarvonshire, that during the
twentieth century have usually been taken as 'Pre-Cambrian'. This
latter assignment is of very doubtful propriety. The metamorphic
rocks of Anglesey, together with the similar rocks in south-west
Caernarvonshire, have always been properly placed below the
Cambrian.

We are here concerned with the documentation of the results of
research and we are at once confronted with the remarkable and
unfortunate fact that Sedgwick allowed his work to go unrecorded
except for the casual and imperfect reports of the oral accounts he
gave at meetings of the Geological Society, the Cambridge Philosophical Society, and the British Association for the Advancement
of Science, during the period 1831 to 1838 [127]. Meanwhile
Murchison was giving brief but careful summaries in the *Proceedings of the Geological Society* of the stratigraphical succession he
was working out, and finally in *The Silurian system,* 1839 [141], he
produced what is probably the greatest work ever to be published
embodying the results of a single piece of original research by one
man. This work detailed and established practically the whole succession of the stratigraphical formations and their fossil-contents
(and associated igneous rocks) of what we now know as the
Ordovician and Silurian systems, in their type areas.

Murchison had included both the Caradoc and Llandovery series
in his one 'Caradoc' series. Sedgwick, 1853 [168] recognised the distinction, which was confirmed by Salter and Aveline, 1854 [170].

The history of the making of this distinction has been given by Jones (1921).

Murchison published the first edition of his *Siluria* in 1854 [169]. This incorporated much of *The Silurian system* of 1839 but surveyed those ever-widening regions which he was including in his 'Silurian' domain. In 1855 Sedgwick put forward an elaborate defence of his Cambrian system, but the most important part of this large volume *Palaeozoic rocks and fossils* [173] was the description of the fossils by Frederick M'Coy.

We are not here concerned with the regrettable and long drawn-out controversy over what was to be included in Sedgwick's Cambrian and what in Murchison's Silurian. The question should have been finally settled, without further argument, by the establishment by Charles Lapworth, in 1879 [240], of the Ordovician system for a natural middle portion of the Lower Palaeozoic succession. Unfortunately the champions of Sedgwick would not let the matter rest after his death. A brief review of this controversy has been given by the present writer (1969). Details of the work of Sedgwick and Murchison will be found in the well-known biographies by Clark and Hughes (1890) and Geikie (1875) respectively.

37 Carboniferous and Cretaceous

In addition to Murchison's *The Silurian system*, the second half of this decade of the eighteen-thirties is notable for two other works which established the rock-succession, lithology, palaeontology, and structure of key areas. These also were the result of many years' work, and the three works together make a complementary trilogy as they deal with well-separated regions, geologically and geographically. They may be said to be analagous to the three works of some twenty years earlier—Farey (1811), Webster (1816), and Otley (1823)—which were mentioned as constituting a somewhat similar kind of set (theme 30). The great difference in magnitude, detail, and grasp between this later trio and the earlier one is a measure of the enormous general advance made in geological knowledge during the two intervening decades.

John Phillips's book on the Carboniferous Limestone district of Yorkshire, 1836 [135] was the second part of his work on the county. The first, on the eastern, Mesozoic part (1829) we have already mentioned (theme 33). These two quarto volumes have been an inspiration to all later geologists working in Yorkshire. In the second volume he established and described, in particular, his 'Yoredale series'. This is a facies of the uppermost zone (D2) of the Lower Carboniferous (Viséan), its highest part corresponding perhaps with the lowest part of the Upper Carboniferous (Namurian). The volume is also especially well-known, and frequently referred to, for its many figures and short descriptions of Carboniferous Limestone fossils. As in the case of the first volume, many of these were newly recognised species.

William Henry Fitton's work, 1836 [134] on the Cretaceous and upper part of the Jurassic in southern England is a paper of 386 pages in the *Transactions* of the Geological Society. It embodied the work of more than twelve years, some important preliminary findings having been published in the *Annals of Philosophy* in 1824. This part of the stratigraphical succession had by no means been neglected previously (in contrast to Murchison's Silurian region), but there were puzzling features of the Cretaceous succession, particularly as regards the 'Greensand' and 'Ironsand' formations, and no great detail had hitherto been recorded. Fitton definitely established the truth and particulars of these matters, with many beautifully drawn maps, sections, and plates of fossils.

38 Further geological treatises 1851-1903

We have already considered some of the discussions of general principles that were written during the second decade of the nineteenth century (theme 22) and the vastly fuller and more correct expositions and appreciations by Lyell, De la Beche, and Phillips in the fourth decade (theme 35). Under the present heading we mention some outstanding general treatises written during the second half of the nineteenth century all of which deal very largely with the geology of Britain.

De la Beche's *Geological observer*, 1851 [162] is a thorough and most attractive exposition of general geology, with examples of the significant appearances and phenomena at home and abroad. It is not just a practical guide to field-work, as its title might suggest.

John Phillips's *Manual of geology*, 1855 [172] and J. Beete Jukes's *Student's manual of geology*, 1857 [176] were standard works for some twenty years.

The general works of Archibald Geikie may be examined a little more closely, although we list in our annals only two of them. While pursuing his researches in particular areas and on particular subjects Geikie kept before him the whole panorama of geological science and produced a series of incomparable textbooks. The first was a largely re-written edition (1872) of Jukes's *Manual*. Then came two very small 'primers', *Physical geography* and *Geology* (1873). The author tells us that 'the composition of those two little books was one of the most difficult tasks I ever undertook.' His efforts were rewarded for they had an enormous success, immediate and continuing, and were translated into most European languages. The *Field geology*, 1876 [224], the *Class-book of physical geography* (1877), and the *Class-book of geology* (1886) came in succession. All these works went into many editions. In 1882 appeared the first edition of the *Text-book of geology* [249] which, in its successive editions of 1885, 1893, and 1903 served as a standard work which may be said to have carried on the function of Lyell's *Principles* and *Elements*. It was based in the first place on a long article contributed to the *Encyclopaedia Britannica* in 1879. A storehouse of leading geological facts, a mine for the extraction of clear definitions and explanations, an authoritative exposition of established principles and, with its copious notes and references, a world-wide guide to the records of discovery, the two-volume edition of 1903 is still indispensable. As far as works produced by British geologists are concerned, it is the last of the great general 'Geologies', later works being either shorter and more elementary or being treatises on special branches of the science.

Phillips's *Manual* was edited, greatly enlarged, and published in two thick volumes by H. G. Seeley and R. Etheridge in 1885 [261], and in 1886 and 1888 were published the two splendidly produced volumes of Joseph Prestwich's *Geology: chemical, physical, and stratigraphical* [264].

A. H. Green's *Physical geology* [225] first published in 1876 was for long the standard work on the subject, later geologists paying tribute to it as a treasured storehouse and guide.

Three general works on British geology were published during the seventies and eighties. H. B. Woodward's *Geology of England and Wales* [227], in its second edition of 1887, is, to the present writer at least, the most attractive book on the subject that has ever been produced. Professor Marr used to advise his students to get hold of a copy (then, as now, out of print), whatever more up to date works they might buy. A. J. Jukes-Browne's two books have always been very well-known, particularly in the editions published shortly before the author's death in 1914. The *Stratigraphical geology*, 1884, 1912 [256] contains useful comparisons with foreign stratigraphy and an up-to-date work on just these lines has long been a desideratum in geological literature. *The building of the British Isles*, 1888, 1911, 1922 [270] was the first attempt at a careful analysis of fact and inference respecting the history of the palaeo-geography of Britain (see theme 67).

39 Sixty years of palaeontology

From among the very large amount of literature on British fossils published during approximately the six decades bringing us up to the end of the century we have selected works which illustrate the main branches and aspects of palaeontology, grouped as follows :

A review of previous knowledge. Buckland, 1836 [133].

Buckland's 'Bridgewater treatise' on *Geology and mineralogy* put forward the author's evidence of a divine purpose in the organic and inorganic history of the earth. It gives an excellent account of British palaeontology as known up to that time.

Catalogues of fossils. Morris, 1845 [151], Etheridge, 1888 [267].

We quote from the Preface to the second of these works, as showing the growth of discovery and record of British fossils up to the date of its publication :

'The preparation of the MS. of this *Catalogue of the fossils of the British Islands stratigraphically and zoologically arranged* was commenced in 1865. At the outset it was in-tended merely to facilitate my own work as Palaeontologist to the Geological Survey of Great Britain. . . . In the year 1822 only 752 extinct species of all classes in the Animal and Vegetable Kingdom were known and described. In 1854, 1280 genera and 4000 species were catalogued by Professor J. Morris; at the close of the year 1874 no less than 13,300 species had been described and for the most part figured; now 3750 genera and 18,000 species comprise the census of the British Fossil Fauna and Flora, all of which have been re-corded in Monographs and serial works dealing with British Geology and Palaeontology. The present volume is devoted to the complete analysis of the Palaeozoic species only, ranging from the Cambrian to the close of the Permian deposits. They comprise altogether 1588 genera and 6022 species arranged stratigraphically (or in the order of time), and also classified zoologically.'

Fossil fruits and insects. Bowerbank, 1840 [142], Brodie, 1845 [150].

Two early works on special kinds and groups of fossils : the first on the fossil fruits of the London Clay, the second on insects from English Mesozoic rocks.

Early monographs of the Palaeontographical Society. Wood, 1848-82 [156]; Milne-Edwards and Haime, 1850-5 [159]; Jones, 1850-99 [158]; Davidson, 1851-86 [161]; Owen, 1851-89 [163]; Wright, 1857-80 [198]; Salter, 1864-83 [197].

The Palaeontographical Society began publishing its long series of monographs in 1848 with the first part of *The Crag Mollusca* by Searles V. Wood the elder. Very soon after were published the first separate monographs of what in effect constituted whole series : T. Rupert Jones's on Entomostraca and Richard Owen's famous series on Mesozoic reptiles (the smallest and the largest of animals). Mention of the remaining monographs brings vividly to mind the plates of illustrations (in particular) so often having to be consulted : Milne-Edwards and Jules Haime on corals, Davidson's massive volumes on the brachiopods, Wright on echinoids, and Salter on trilobites.

'Figures and descriptions of British organic remains' issued as *Memoirs of the Geological Survey* between 1849 and 1872

They were grouped in so-called 'Decades', each containing ten very full descriptions with beautifully engraved plates of particular species. We list the first four decades, 1849-52 [157] which were prepared by Edward Forbes, the Survey palaeontologist, and which dealt with echinoderms and trilobites.

The later works of the century, or begun during the century, we shall enumerate under a few main headings.

INVERTEBRATE PALAEONTOLOGY

Hinde, 1884 [225] and 1887 [265], fossil sponges; Hudleston, 1887-96 [266], Gasteropoda of the Inferior Oolite; Whidborne, 1889-1907 [279], the Devonian fauna of the south of England; Lapworth, 1891 [292], the Lower Cambrian trilobite, *Olenellus callavei* (see also theme 50); Hind, 1894-6 [305] and 1896-1905 [314], on Carboniferous lamellibranchs; and Woods, 1899-1913 [332], on Cretaceous lamellibranchs.

VERTEBRATE PALAEONTOLOGY

Prestwich, 1864 [196], Mammalia and flint implements in the Drift; Dawkins, 1866-72 [202] and 1874 [214], Pleistocene mammals; Traquair, Powrie, and Lankester, 1868-1914 [205], fishes of the Old Red Sandstone; Newton, 1878 [237], Cretaceous fishes, 1888 [273], a very remarkable description of the organs of the skull of a pterosaurian, 1893-4 [301], reptiles from the Elgin Sandstone (Permo-Triassic), and 1894 [309], Pleistocene vertebrates from a

H

fissure in Kent, and other memoirs; Woodward and Sherborn, 1890 [284], a catalogue of British fossil vertebrates; and Traquair, 1899-1905 [330], Silurian fishes of South Scotland.

EARLY MAN

The three well-known books: Lyell's *Geological evidences for the antiquity of man,* 1863 [192]; Evans's *Ancient stone-implements, weapons, and ornaments of Great Britain,* 1872 [209]; and Dawkins's *Early man in Britain,* 1880 [245]. We may also include here the later book, Sollas's *Ancient hunters,* 1911 [389].

PALAEOBOTANY

Kidston, 1894 [306], Carboniferous plants and the stratigraphical divisions based on their vertical distribution; Seward, 1894-5 [310], Wealden flora, 1898-1919 [324], the standard several-volume work on fossil plants.

40 The Ice Age

The history of the 'glacial theory' and of the circumstances of its arrival and acceptance in Britain in 1840 have been detailed in several works, notably in those of North (1943), Chorley and others (1964), Davies (1969) and White (1970). The British evidence was not documented by any special description of the phenomena till Ramsay's paper, 1852 [166] on the drifts and surface markings of North Wales, which was followed in 1860 by *The old glaciers of Switzerland and North Wales* [184], surely the smallest book (and one of the most attractive) in the whole annals of British geology. In 1858 appeared in the *Quarterly Journal* of the Geological Society the first of T. F. Jamieson's long list of papers [179, 183, 191, 201, 216, 361], covering a period of 48 years, on the glacial deposits of northern Scotland, and in 1863 Archibald Geikie's on the Scottish glacial drifts [190], the first paper in the first volume of the Glasgow Geological Society's *Transactions.* It was Archibald's brother, James Geikie, who took over the family interest in Pleistocene glaciology, and the first edition of his *Great ice age,* 1874 [215] was a landmark. Newbiggin and Flett (1917) tell us much of interest about this book. Peach and Horne, taking the latest as well as the earliest period in the geological history of Scotland as one of their intensive fields of study, wrote papers on the glaciation of the most northerly parts of that country, 1879 [242], 1880 [246], 1881 [248]. Lamplugh, 1891 [291] described in detail the drift deposits of Flamborough Head; and in 1894 appeared, twenty years after the *Great ice age,* H. Carvill Lewis's *Glacial geology of Great Britain and Ireland* [308], the next landmark in the establishment of our growing knowledge.

41 Internal structures of rock-bodies: slaty cleavage

In 1835 was read before the Geological Society, and published immediately, Sedgwick's most important written treatise : a very careful examination of the internal structural features of rock-bodies, particularly slaty cleavage, stratification, and jointing [131]. Fifty years later Harker, 1885 [257] gave his paper on slaty cleavage before the British Association, in which the early history of the 'cleavage problem' was summarised. The whole matter has been reviewed, in its theoretical aspects, by Wilson (1946).

42 The older Tertiary formations of south-east England: water supply

Joseph Prestwich, in a series of papers from 1846 to 1857 and a supplement of 1888, and particularly in three papers published in the Geological Society's *Quarterly Journal* between 1850 and 1854 [160], elucidated the hitherto little-known stratigraphy of the beds in England lying between the Chalk below and the main part of the London Clay above and correlated them with the corresponding beds which had been described in France and Belgium. He called these beds the Lower London Tertiaries and divided them into the Thanet Sands, the Woolwich and Reading series, and the basement beds of the London Clay (later separated from that formation and called the Blackheath and Oldhaven beds). This important part of the stratigraphical succession was thus established and the author became an acknowledged leader in the Tertiary geology of Europe. (See also Buckland, 1817, in our theme 24).

Prestwich's *Geological inquiry respecting water-bearing strata of the country around London,* 1851 [164] was the first important work on the conditions of water supply. It was nearly fifty years later that the Geological Survey began publishing memoirs exclusively devoted to water-supply, and to these the most active and enthusiastic contributor was William Whitaker, 1899-1921 [331]. The investigation of this vital aspect of economic geology continues to be one of the main concerns of the Survey, and its work in this connection has been recorded by the two historians of this institution, Flett (1937) and Bailey (1952).

Prestwich's geological work has been summarised by Geikie in a section appended to the *Life and letters of Sir Joseph Prestwich* edited by Lady Prestwich (1899).

43 *Volcanoes and their products in Scotland*

This subject is so vast, established in such well-known original works, and already summarised and historically reviewed so completely and authoritatively, that all we do here is to put down the titles of these researches and writings, with the minimum of comment.

There are three main geological periods of volcanic activity—Devonian (Old Red Sandstone), Carboniferous (with some Permian), Tertiary—and up to the present day, three main periods of investigation. The first of the latter, taken to constitute the present theme, occupies much of the nineteenth century and is associated chiefly with the names of J. W. Judd and Archibald Geikie, particularly with Geikie. The following works are in our selection :

Maclaren, 1839 [140]. Fife and the Lothians. Old Red Sandstone and Carboniferous

Geikie (in Geikie and Howell), 1861 [185]. Edinburgh. Carboniferous. (With revision by others in the 1910 edition)

Geikie, 1861 [186]. General but largely Old Red Sandstone

Judd, 1873-93 [212, 258, 263, 277, 281, 299, 300]. Tertiary

Geikie, 1879 [239]. Carboniferous

Geikie, 1888 [268]. Tertiary

Geikie, 1896 [313]. Tertiary

Geikie, 1897 [318]. General (in *Ancient volcanoes*)

There is a full documentation and discussion of the history of research during this period in the last of these items.

44 *Conditions of formation of sedimentary rocks*

The early geologists, producing their large works on the stratigraphical geology and stratigraphical palaeontology of their regions, well realised that one of the main objects of collecting and recording all the facts was to enable views to be drawn of the physiographical and biological conditions which prevailed at the time and under which the several strata were laid down. Thus we find Phillips in his Yorkshire, 1836 [135] discussing the 'circumstances attending the deposition of the Mountain Limestone formation' and Fitton in the same year, in his long paper on the Cretaceous [134], having a section on 'Geology of the Wealden' treating the same question. Further back in our annals, we have seen how certain Carboniferous and Tertiary strata were being diagnosed as of marine or freshwater origin by the comparison of their fossil shells with modern forms characterising each of those conditions (themes 4, 26). However it was De la Beche who wrote what seems to have been the first extended treatise, 'On the formation of the rocks of South Wales and south-western England'; the first memoir in the first volume of the Memoirs of the Geological Survey, 1846 [154]. This matter has never been lost sight of, and we have had for instance, Marr's 'Conditions of deposition of the Stockdale Shales', 1925 [463] and his well-known small volume The deposition of the sedimentary rocks (1929). In recent years there has been a renewed interest in the physical and biological conditions which are to be inferred from the character and distribution of the rocks and their contained fossils. This results from the very detailed and exact study of the rocks and their sedimentary structures and of the fossils and faunas, with the elaboration of new methods and with a growing knowledge of the conditions and processes of today.

45 Geological Survey: South-west England: Devonian and Old Red Sandstone

One of the most important events in the history of British geology was the establishing of the Geological Survey in 1835 with De la Beche at its head. The history of this national institution has been written by Flett (1937) and Bailey (1952), and there is much about it in the biographies of its officers, particularly those of De la Beche's famous successors as director-general: Murchison (Geikie, 1875), Ramsay (Geikie, 1895), and Geikie himself (Geikie, 1924). In 1839 [139] appeared the first memoir (called 'Report'), to start the great series which, with the accompanying maps, has ever since been the chief source of systematic and detailed information about prescribed areas in Britain. This first memoir covers a large region which was later parcelled out in a number of separate memoirs. The latest memoir and map of a particular area may be anything up to a hundred years or more old and be awaiting revision. De la Beche's thick volume of 648 pages and folding map, sections, and plans, dealt with the very complicated geological region of south-

west England, one where the rock-formations were not directly com-
parable with those known in other parts of Britain at the time.
'Greywacke' was still the name used for any of the older (Palaeozoic)
rocks not manifestly Carboniferous or Old Red Sandstone. Therefore the several rock-groups were not properly sorted out. The
Report is especially valuable, even at the present day, for the very
full description of the mineral lodes.

It was the researches, just a little later, of Sedgwick and Murchison, with the very material help of William Lonsdale on the age
of the fossils, that put the succession right and which resulted in the
establishing of a new geological system—the Devonian—founded
on a normal marine succession of deposits. Papers were read to the
British Association in 1836 and to the Geological Society in 1837
and 1839 and the full account was published in 1840 [144]. The
whole story is told at length in Geikie's *Life of Murchison* (1875)
and more concisely in his *Founders of geology* (1905). Meanwhile
John Phillips had been examining fossils from the region and the
results were published in 1841 [146] as a supplementary volume
to De la Beche's *Report,* but this work was somewhat undercut by
the finds, descriptions, and interpretations of the fossils in the
researches of Sedgwick, Murchison, and Lonsdale. The term
'Palaeozoic' appears in the title of Phillips's book; it had been in
use for a year or two, but Phillips here uses it in exactly its modern
sense, and he is the first to use the term 'Lower Palaeozoic' and, at
once, in its modern sense.

Godwin-Austen's paper on south-east Devon in the *Transactions,*
1842 [148] combines the substance of five discourses read before the
Geological Society between 1834 and 1840. The author (who was
without the 'Godwin' in his name at that time) refers to the earlier
of Sedgwick and Murchison's announcements (1836) and to De la
Beche's *Report* (1839), but he seems to have arrived at his results
independently. Dewey in the *Regional geology: South-west England,*
1948 [535] goes so far as to say that 'this paper may be regarded
as the foundation on which all subsequent work in the area was
built.' Austen does not use the name 'Devonian'.

Pengelly's paper on the south-west coast of Devon, 1855 [171] was
published in the *Transactions* of one of the oldest of all geological
societies, the Royal Geological Society of Cornwall.

The Devonian system, proposed for the rocks of south-west
England and, a few years later found, by Sedgwick and Murchison
themselves, to be more completely developed on the Continent,
occupies the same position in the stratigraphical sequence as the
very different Old Red Sandstone. This latter name was introduced
by Jameson, 1805 [57] as a translation of Werner's *Aelter rother*

Sandstein with which he correlated it, but Werner's formation was, on the whole, Permian in age. The British Old Red Sandstone was, as we have seen, first tabulated in its proper position by Buckland, 1818 [89] but was for twenty years classed with the Carboniferous system. It was Murchison who first called it a 'system' on its own, 'in order to convey a just conception of its importance in the natural succession of rocks' (*The Silurian system*, 1839 [141]). The 'Old Red Sandstone' is no longer used as a system-name but a special development (facies) of the Devonian system.

Hugh Miller's *Old Red Sandstone*, 1841 [145] is a classic of geological literature of an unusual kind. It makes some original observations on these rocks in the north of Scotland and describes his finds of fossil fish, with remarkable 'restorations'. Its significance in the progress of geology is that, by the charm of its personal narrative and description, it awoke a widespread interest in the methods and results of geological inquiry. His work and influence have been discussed in one of Archibald Geikie's illuminating essays (1905 (b). Mention should here be made of the activities and discoveries of Robert Dick, the baker of Thurso (Smiles, 1878).

Nearly forty years later (1878) the other classic work on the Old Red Sandstone was written : Geikie's long paper 'On the Old Red Sanstone of western Europe' in the Royal Society of Edinburgh's *Transactions*, 1878 [234]. This discussed the stratigraphical classification, original areal extent, and conditions of formation of these rocks.

46 The Malvern Hills

The Malvern Hills form one of the most striking districts in Britain, both scenically and geologically. It was first investigated by Horner, 1811 [67] and later by Murchison, 1839 [141]. Phillips's large memoir of 1848 [155], one of the earliest to be published by the Geological Survey, is one of those fundamental works which at once establish our knowledge of a region and which have never been superseded, though many new facts are subsequently recorded. The Malvern ridge itself is composed chiefly of a variety of Pre-Cambrian gneisses (Malvernian); the lower hills on the west of folded Silurian, and the plain on the east of flat-lying Trias. All through the period of investigation, questions as to the geological history of the region have been obviously challenging, in particular those as to the natures of the contacts between the Malvernian and the Silurian on the one hand and between the Malvernian and the

Trias on the other. The possibilities here as to each of these contacts are: (1) undisturbed unconformity, (2) faulting affecting the sedimentary formation (Silurian, Trias) as a whole and thus being later than this formation, (3) unconformable deposition upon and against a fairly steep surface produced by contemporaneous faulting, and (4) an unconformable surface which has later become tilted or folded. Horner had considered the Malvernian to be an igneous mass intruded in post-Triassic times. Murchison considered that both the junctious were unconformities. Phillips found unconformity on the west (Silurian) and postulated unconformity on the east (Trias), the old Triassic shoreline being, in his opinion, produced by a pre-Triassic fault. On the Survey one-inch maps of 1855 a fault is boldly drawn along the line of the Malvernian-Trias boundary, with the obvious implication that the fault is of post-Triassic age. Groom, in his series of papers at the turn of the century [326, 334, 342, 347] summarised in *Geology in the field,* 1910 [376]), gave faulting as the nature of the contact on each side of the Malvernian ridge. It should be mentioned that the folds in the Silurian rocks were caused by pressures which, because the Carboniferous rocks in neighbouring areas were found to be affected, must have been post Carboniferous in age.

It seems that faulting (in our sense 2 above) has been altogether too readily postulated, before the logically prior consideration of simple unconformity has been given full weight, as the Silurian and the Trias would each of them be naturally unconformable to the Pre-Cambrian. While detailed discussion has been largely concerned with the western, Silurian boundary, the eastern, Trias boundary has been comparatively neglected. Here a major fault has always been shown on the maps as one of the main lines of fracture in the structure of Britain as if there were no question about it. But Falcon (1947) has shown that there is indeed a great question about it, and his arguments demand close attention. Further commentaries, with bibliographies, records, and theories have been given by Raw (1952), Butcher (1962), in articles and correspondence in the *Geological Magazine* (1961-5), by Phipps and Reeve (1967, 1969), Jones and others (1969), and Brooks (1970). This shows how very much alive the problem still is.

Cambrian rocks occur at the south-west corner of the Malvern Hills. There is also confusion here in the interpretation of the contact of these rocks with the Pre-Cambrian. Faults are envisaged by some, particularly Groom, but there seems no reason why the contact should not be essentially a simple unconformity. The Silurian here overlies the Cambrian; there appears to be little structural discordance of any kind (both Phillips and Groom show a perfect concordance), but there must be a large stratigraphical non-sequence as the whole of the Ordovician is absent. It is reasonable to assume that the base of the Silurian is continuous across the outcrop of this Cambrian onto the outcrop of the Pre-Cambrian of the main mass of the Malvern Hills.

The geology of the Malvern district is epitomised in the *Regional geology* handbook on the *Welsh Borderland* [539].

47 Concealed coalfields

'In 1834 Conybeare wrote to the *London and Edinburgh Philosophical Magazine* "On the probable future extension of the coalfields at present worked". He showed that the structure of the coalfields east of the Pennines was such that the Coal Measures passed southwards or eastwards beneath a covering of newer strata. We are now so familiar with concealed coalfield exploration that it is useful to be reminded that someone had to enunciate the principle for the first time' (North, 1956).

The first treatise on the subject was that by Godwin-Austen in 1856 'On the possible extension of the Coal-Measures beneath the south-eastern part of England' [175]. This dealt with a region far removed from any exposed coalfield. The important Geological Survey memoir on the concealed coalfield of Yorkshire and Nottinghamshire, by Gibson, had its first edition in 1926 [470]. One hundred years after Godwin-Austen's paper, the book *Concealed coalfields,* by Wills [633], was published.

48 The earlier days of petrology and petrography

Petrology, the science of the rocks in themselves, comprises two main branches : (1) petrography, the description and study of rock-specimens and rock-types, their mineral composition and texture, and (2) petrogenesis, the origin and mode of formation of rocks, a term used almost exclusively in connection with the igneous rocks, as these questions concerning the sedimentary rocks form essential parts of physical and stratigraphical geology. In practice, particularly in specialised papers or chapters, works with 'petrology' in their titles are usually works on 'petrography'. Ever since the middle of the nineteenth century (the term was hardly ever used before), this has meant the study of rocks in thin section ('slices') under the microscope or an analysis of their component grains. This method was first used in Britain, by Henry Clifton Sorby, and it revolutionised the whole of geology by penetrating into previously hidden secrets of the utmost significance. (See Wilcockson, 1947).

We here list the items from our selection.

Sorby, 1858 [180]. 'On the microscopic structure of crystals, Calcareous Grit of the Yorkshire coast.' A paper of only six pages inaugurated a new era in the investigation of rocks.

Sorby, 1858 [180]. 'On themicroscopical structure of crystals, indicating the origin of minerals and rocks.' An even more momentous publication.

Allport, 1874 [213]. 'On the microscopic structure and composition of British Carboniferous dolerites.' 'Allport was the first to reject the age-criterion in nomenclature of rocks' (Tomkeieff, 1954).

Ward, 1875 [220]. The 'microscopic rock-structure' of modern volcanic rocks compared with the 'ancient volcanic rocks' of Wales and the Lake District. The history of the application of the method to date is reviewed.

Ward, 1875-6 [221]. Two papers on the intrusive igneous and associated metamorphic rocks of the Lake District, the facts being considered by the authorities at the time as too technically petrographical for inclusion in Ward's contemporary Survey memoir.

Allport, 1876 [222]. 'On the metamorphic rocks surrounding the Lands-End mass of granite.'

Allport, 1877 [228]. 'On certain ancient devitrified pitchstones and perlites' in Shropshire.

'The principal object of the present communication is twofold : in the first place, to bring under the notice of the Society the occurrence in Shropshire of an extremely interesting series of ancient vitreous and semivitreous lavas, and their associated agglomerates and ashes; and in the second, to show, from an examination of their structure and composition, that originally they were absolutely identical with some of the glassy volcanic rocks ejected during the most recent geological periods.'

Rutley, 1885 [262]. Geological Survey memoir on the rhyolitic ('felsitic') lavas of England and Wales. Rutley had contributed to several district memoirs and here describes particularly these Lower Palaeozoic lavas of Wales and the Lake District.

Teall, 1888 [274]. The famous *British petrography*. This large quarto-sized book summarises and illustrates with coloured drawings of microscope slides the knowledge of British petrography to date, Teall himself having greatly contributed to the matter.

49 Geological maps of England and Wales

We have already noticed (theme 27) the early maps of Smith and Greenough. No maps of the whole country on such a large scale as theirs have ever been issued since. We here notice two general maps on about half the scale. The first is Ramsay's of 1859 [182]. In 1895 Archibald Geikie wrote : 'Ramsay at this time condensed the information on the published Survey maps, and produced a geological map of England and Wales on the scale of twelve miles to an inch, which is still the most serviceable general map of the kingdom.' This was not published by the Survey itself, nor was that prepared by Geikie, soon after writing the above words, in 1897 [319]. Geikie's map was the best available for the next fifty years and still remains the clearest and most attractive to the eye.

We are not recording in our annals the Geological Survey maps, on the scales of Quarter-inch, One-inch, and Six-inches to the mile, which form the original field surveys and primary reductions. A separate detailed study of the successive issues of the maps on these scales would itself provide a strong thread through the pattern of the history of British geology since 1844, when the first map (One-inch, part of South Wales) was published.

As for other maps (of various or all periods) reference may be made in particular to the account of John Phillips's maps by Douglas and Edmonds (1950), to the comprehensive accounts and bibliographies of maps of Wales and neighbouring parts by North (1928) and Bassett (1967).

50 Various English regions

Jukes, 1853 [167]. South Staffordshire coalfield
Pengelly and Heer, 1862 [189]. Bovey Tracey beds of Devon
A kind of geological formation unique in Britain. Lacustrine deposits of clays and lignites, the latter containing many kinds of plants, including the genus *Sequoia*. As the species are long-range the exact age is uncertain, but an Oligocene age is now assigned. This paper is outstanding in the history of investigation. Summaries

(with references) in Woodward, 1887 [227], Boswell in the *Handbook* of Evans and Stubblefield, 1929 [494], and the *Regional geology* handbook on *South-west England* 3rd edition, 1969 [535].

Morton, 1863 [193]. Country round Liverpool

There was a second edition in 1897, 'virtually a new and larger work'. Between these two dates the Geological Survey had examined the district. There is much about the *Chirotherium* footprints, the animal, evidently a reptile, known only from these 'trace-fossils' (see Tresize (1969), for a summary).

Phillips, 1871 [208]. Oxford and the valley of the Thames.

> 'Phillips used a broad canvas, taking in an area from the Malvern Hills to London and almost to Cambridge "embracing the whole period of geological time from the oldest rocks of Malvern to the latest prehistoric alluvium". To attempt such a work now, commensurate with the state of knowledge at the present day, would take three volumes—and three lifetimes' (Arkell, *Geology of Oxford,* 1947).

The book is specially rich in palaeontology.

Whitaker, 1872 [210], 1889 [280]. His important memoirs on the London Basin

For comments on these and the later accounts of the Capital region see the British *Regional geology* handbook on *London and Thames valley,* 3rd edition 1962 [544].

Judd, 1875 [218]. Rutland

In the earlier part of his career Judd worked on the Geological Survey and an outcome of this was his memoir on the Jurassic oolites of Rutland.

Topley, 1875 [219]. The Weald

One of the most important of the earlier Geological Survey memoirs. Topley describes this famous British geological region, one of the best-known but one which always presents new problems of detail, and gives a full account of the growth of knowledge. The latest edition (4th) of the *Regional geology handbook, The Wealden district* [537] also does justice to the science of geology by giving a full historical summary.

Green, 1869 [206], 1878 [235]

During Green's period on the Geological Survey his most important work was the production, with collaborators, of the memoir on North Derbyshire in 1869 (2nd edition 1887) and of that on the Yorkshire coalfield in 1878.

Callaway, 1877 [232], 1878 [233], 1879-82 [238], 1891 [287]

This series of papers in the *Quarterly Journal* of the Geological Society was the first to establish the occurrence and succession of the rock-groups in that highly interesting geological region, South Shropshire. The discovery of the Shineton Shales of Tremadocian age was announced in 1874 and the full account with description of the fossils was given in 1877. In the 1878 paper Callaway described, in particular, the Lower Cambrian quartzites. In two papers of 1879 and 1882 he described the Pre-Cambrian volcanic rocks of the Wrekin and the Church Stretton hills and those to the west of the Longmynd with petrographical notes added by Bonney. In 1891 he reviewed the Uriconian and Longmyndian rocks, continuing to take both as being Pre-Cambrian in age, considering in particular the kind of break between them and their relative age, problems still remaining today.

Bonney and Hill, 1877-80 [231], 1891 [286]. Pre-Cambrian of Charnwood Forest, Leicestershire

Bonney, 1877 [230], 1883 [251], 1891 [285]. The serpentines and associated rocks of the Lizard district

Bonney, 1884 [254]. A general review of the Pre-Cambrian geology of Britain

Lapworth, 1888 [271], 1891 [292]

Two papers in the *Geological Magazine* announcing the discovery of the *Olenellus* fauna in Britain at Comley in South Shropshire and describing it. This had already been found to indicate a Lower Cambrian age in North America and the Baltic region.

Lapworth and Watts, 1894 [307]

The well-known account of the geology of South Shropshire, repeated, substantially unaltered in *Geology in the field,* 1909-10 [376].

Lapworth, with contributions by Watts and Harrison, 1898 [323]

The geology of the 'Birmingham district', a wide-ranging account including at its four corners the Wrekin, Charnwood Forest, Rugby, and the Abberley Hills.

51 *Central and North Wales*

Apart from the generalities connected with the Cambrian-Silurian controversy (theme *36*), research into the geology of Central Wales was started by the Geological Survey. Andrew Ramsay was in charge of this work, and in 1842 he discovered that the rocks to the north-west of the 'Silurian' region were not, as had been thought, older than the Llandeilo series, but were repetitions, by folding, of formations belonging to Murchison's 'Silurian' system (which included the Llandeilo as its lowest series). Thenceforward the progress of our knowledge of this region is recorded in the maps and sections of the Geological Survey. The first of these were several so-called 'horizontal sections' published in 1845 [152], and a most remarkable fact is that these show the structure of the rocks more accurately and in more detail than anything that has been produced since. Indeed there never has been a detailed account of the heart of the area (Cardiganshire and neighbouring parts) by the Survey or anyone else and, except in bits, it is still something of a geological wilderness, notwithstanding the bird's-eye view given by O. T. Jones, 1912 [394]. This history, and the present position, has been fully reviewed by the present writer (1951 and 1969). The Survey is, however, now again on the ground.

The story in North Wales is a different one for the Geological Survey, as it progressed northwards, became more concentrated in its researches. Eventually the great memoir on *The geology of North Wales* [203] was published in 1866 with a second edition in 1881, at which time it was the most important Survey memoir in existence.

In 1889 was published one of the smaller classics of British geology, Alfred Harker's *Bala volcanic series of Caernarvonshire* [276]. This is both a field study and a petrographical account, and for the latter, references are to particular rock-slices preserved in the Sedgwick Museum at Cambridge instead of to drawings in the text, a surer method, if not so immediately illustrative.

Meanwhile in the extreme south-west corner of the Welsh Lower Palaeozoic tract, Henry Hicks was investigating the old rocks of the St. David's peninsula, 1871-1886 [207, 211]. He advocated a Pre-Cambrian age for certain igneous rocks and established a faunal succession in the overlying Cambrian, except for the lower part. The Geological Survey under Ramsay had mapped the area roughly (maps issued in 1845 and 1857), interpreting the above-mentioned igneous rocks as post-Cambrian intrusives, an interpretation supported by Archibald Geikie during the progress of publication of Hicks's series of papers. Hicks's views were summarised in the *Popular Science Review* in 1881. His assignment of the igneous rocks to the Pre-Cambrian appeared to be confirmed by Green's work on the St David's area, 1908 [368] and has been adopted ever since, but it cannot be said that the matter is settled. The rocks may be generally equivalent to those in north-west Caernarvonshire about which there are doubts regarding their appropriate placing (theme *36*).

Another contemporary paper on the Lower Palaeozoics of Pembrokeshire was that by Marr and Roberts on the Haverford-west neighbourhood, 1885 [260].

Historical notes and comments will be found in Thomas and Jones's paper on the Hayscastle district (1912) and in the Pre-Cambrian and Cambrian chapters of Jukes-Browne's *Stratigraphical geology*, 1912 [256].

52 Breaks in the succession of strata

The question of continuity in a succession of strata is probably the most important one in stratigraphy. It is one that affects successions on a large scale over a wide area and local small-scale successions. As regards the latter, Jukes in the 1862 edition of his *Manual* [176] has the following:

> 'It is quite impossible in any quarry to say of one bed that rests directly on another, that it was not only the next formed bed at that particular place, but that no other bed, or other set of beds, was formed anywhere else in the interval between them. If he place his finger on the plane of stratification between two beds, that little space may mark the lapse of years, centuries or milleniums.'

Large-scale breaks formed the subjects of Ramsay's two presidential addresses to the Geological Society, taking the Palaeozoic in 1863 and the Mesozoic in 1864 [194]. Small-scale breaks, in reference to the Lias, were considered in detail by Walford (1902) and by Trueman (1923). The whole question has recently been discussed, comprehensively and concisely, by Donovan (1966).

53 Geomorphology

There are two books which, in their successive editions, dominated the geomorphology bookshelf till the end of the century: Ramsay's *Physical geology and geography of Great Britain*, 1863 [195] and A. Geikie's *Scenery of Scotland*, 1865 [200]. These were joined by J. Geikie's *Earth sculpture* in 1898 [321]. Three very important studies were made in the Weald district of south-east England; important because of their triumphant vindication of the thesis that the present relief of regions other than the coast is due to fluvial, not to marine erosion in the past. The studies were those of Foster and Topley, 1865 [199], Whitaker, 1867 [204], and Topley in his Geological Survey memoir, 1875 [219]. The whole history of geomorphology up to 1889 when W. M. Davis came on the scene, is treated in *The history of the study of landforms,* by Chorley, Dunn, and Beckinsale (1964) and in *The Earth in decay,* by Davies (1969). These two books deal with the works mentioned under the present heading together with the associated contemporary writings.

Innumerable works on geomorphology, large and small, general and local, have appeared in the present century. Of those of a general nature we have selected two as being particularly attractive and informative : Trueman's *Scenery of England and Wales,* 1938, reissued as a 'Penguin' book in 1949 [564] and Steers's *Coastline of England and Wales,* 1946 [584]. Arber's *Coast scenery of North Devon* [384] is more particularly referred to in theme 72.

54 *The North-West Highlands of Scotland*

'The correct explanation of this structure introduced to geologists a new type of displacement in the earth's crust' (Geikie, 1924). It also placed the whole edifice of the rock-succession in Britain on a firm foundation. This 'correct explanation' was finally established, and in minute detail, by the publication in 1907 of what is probably the most important Geological Survey memoir ever produced, *The geological structure of the north-west Highlands of Scotland,* by B. N. Peach, J. Horne, and others [366]. In that memoir is recounted the history of research, the gradual penetration into the precise truth of the matter through the surrounding mass of uncertainty and error. This history is highly interesting in itself, and it gives an excellent insight into some of the main principles of the science of geology. It is partially told in both the histories of the Geological Survey (Flett, 1937, and Bailey, 1952) and in Greenly's *Memories* (1938); but whereas the history in the memoir is so particularised as to be difficult to follow in outline, the accounts in the three books are rather scrappy. The following are the chief steps in the advance of true knowledge :

'In Sutherland and Ross, Macculloch, between 1814 and 1824, had described a succession which, to use modern names, may be summarised thus : Lewisian gneiss, covered unconformably by Torridonian sandstone [see theme *23*], overlain by quartz-ites and limestone, which alternate with and are succeeded by gneiss and schist forming the main mass of the Highlands' (Bailey).

Roderick Murchison who became the head of the Survey in

1855, made several visits to the area, particularly in 1859 with Ramsay and in 1860 with Geikie. Murchison, supported by these two geologists (who successively followed him as head of affairs in the British geological world), agreed with Macculloch that the order of observed superposition was in fact the order of decreasing age. This meant that, in Murchison's view, the eastern schists, named the Moine Schists in 1888, were the youngest rock-group (as an originally formed mass). As metamorphic schists they are no doubt the youngest rock-group.

In 1856 Nicol read to the Geological Society a paper, published in 1857 [177] describing the unconformity between the quartzite (together with the succeeding limestones) and the underlying red sandstones; and in 1857 Salter made a communication to the British Association in which he identified fossils, which had been found by Charles Peach in the limestones, as of Ordovician age (incidentally, with North American affinities). Salter thus considered the whole quartzite-limestone sequences to be Ordovician, and the red sandstone, later called Torridon Sandstone, or Torridonian, to be Cambrian.

The main feature of Nicol's second very important paper, 1861 [187] was his demonstration that the Moine Schists were separated from the underlying Ordovician limestones not by a conformable contact but by a large dislocation, which he interpreted as a steeply-inclined fault or fault belt. The eastern gneisses and schists, (the Moine) Schists were, in his view, the Lewisian gneisses and schists repeated as a result of this dislocation.

Callaway in 1883 [252], confirmed Nicol's dislocation, but was not definite as to its character, particularly its inclination. He considered the eastern schists to be 'Archaean' (Pre-Torridonian, i.e. Pre-Cambrian) but, recognising the difference in character between them and the Lewisian, he placed them as a separate group, younger than the Lewisian.

The great discovery that the dislocation separating the Moine Schists from the underlying rocks was a low-angled fault, an overthrust, was made by Lapworth, 1883 [253], and 1885 [259]. He left open the question as to what the metamorphic eastern schists were before they became metamorphosed.

Geikie, now head of the whole Geological Survey, realised that the succession and structure of this most important region required a detailed official investigation, and in 1883 he 'dispatched his two best field geologists, Peach and Horne, to Durness' (Bailey). Geikie expected that they would confirm Murchison's view which he, as a young disciple, had supported, but it turned out quite otherwise, and the truth contained in the successive interpretations of Nicol,

I

Callaway, and Lapworth was extracted and amplified. Geikie immediately announced the new official findings, which were for him indeed a triumph, not a 'disaster', as Bailey calls it. The important 'Report' was published in the *Quarterly Journal* of the Geological Society in January 1888 [269]. The nomenclature of the rockgroups became definitely established as we know it today.

But still one discovery of the greatest significance remained to be made. This was the finding of the *Olenellus* fauna in the quartzites, the lower part of the series hitherto taken to be wholly Ordovician by Peach and Horne, 1892 [295]. This fauna characterises the Lower Cambrian. The Torridonian was removed from the Cambrian system and placed as the newer of the two great Pre-Cambrian groups of the North-west Highlands.

The drama, or the first prolonged 'act' of the drama, was closed by the publication of the *North-west Highlands* memoir in 1907 with the mention of which we began this synopsis. Two works of this period which are in our selection but which we have not mentioned are Bonney's petrological notes on rocks from Loch Maree [244] and Clough's memoir on Cowal, southern Argyll [317].

Many works have been published since 1907. The following are in our selection :

Jehu and Craig, 1923-34 [443]. The Outer Hebrides (Lewisian)

Peach and Horne, 1930 [504]. The unfinished book on the geology of Scotland

The chapters deal chiefly with the North-west Highlands.

Sutton and Watson, 1951 [612]. The pre-Torridonian metamorphic history of the Loch Torridon and Scourie areas. Two periods of metamorphism of the Lewisian, the Scourian and the Laxfordian, widely separated in time, are recognised.

Kennedy, 1951 [610]. Sedimentary differentiation and Moine-Torridonian correlation. One of the recent major additions to our understanding of the geology of this region is the gradual growth of the opinion that the Moine Schists are the metamorphosed equivalents of the Torridonian Sandstone.

Sutton and Watson, 1953 [618]. Supposed Lewisian inlier, Ross-shire

McIntyre, 1954 [622]. History of discovery, age, and tectonic significance of the Moine thrust

Sutton and Watson, 1954 [623]. Moines in Ross-shire

Bailey, 1955 [626]. Moine tectonics and metamorphism in Skye

Kennedy, 1955 [629]. Tectonics in Morar (Inverness)

Sutton and Watson, 1959 [640]. Structures in the Glenelg area (Inverness)

Dearnley, 1962-3 [644, 647]. Lewisian of the Outer Hebrides

Finally, of great value are the relevant parts of (1) Phemister's account of the *Northern Highlands* in the latest edition (1960) of the *Regional geology* handbook [552], and (2) the several chapters, by George, Watson, Johnson, and Walton, in the *Geology of Scotland*, 1965 [651].

55 *The main mass of the Highlands*

The modern era in the history of research into the manifold problems of the geology of the Highlands of Scotland may be said to have begun, and in force, with the work of Edward Bailey and his associates. This was inaugurated with the publication in 1909 of the paper on the 'cauldron subsidence' of Glen Coe, followed by other papers in the *Quarterly Journal* of the Geological Society (1910-22) and by the Geological Survey memoir on Ben Nevis and its surroundings in 1916. But the paper by George Barrow in 1893 was a forerunner of this modern era because it established in detail the principle that rocks affected by the 'thermometamorphism' produced by intrusion of an igneous mass show, by the presence of certain key-minerals, the degree to which they have been affected. In any case these several rocks with their characteristic minerals occur in bands, grading one into the next, at successive distances away from the intrusion. Such bands are the outcrops of three-dimensional shells. In a diagrammatic case (which Barrow's hardly was) this results in what afterwards became known as a 'metamorphic aureole'. Barrow's zones of progressive metamorphism are due partly to pressure as well as to heat and are thus to some extent of a 'regional' nature. His area was in Angus near the south-east border of the Highland (Dalradian) mass. Some sixty years later Read (1952 and 1956) described from the Buchan area of Aberdeenshire, to the north-east, a set of metamorphic zones seeming to depend on physical conditions and a time-sequence rather different from those in the area to the south-west. We thus have two 'types' of thermo-dynamic metamorphism which have become known as the 'Barrovian' and the 'Buchan' types. (Barrow's zones are discussed in Turner and Verhoogen's *Igneous and metamorphic petrology*, 1960).

The present state of our knowledge of the geology of the Highlands, resulting from the immense amount of work that has been done during the last sixty years—and it is being very actively pursued today—is summarised and illuminated in the following works : The *Regional geology* handbooks on the *Northern Highlands* (3rd edition, 1960) and the *Grampian Highlands* (3rd edition,

1966); *The British Caledonides* (1963, particularly the chapters by Knill, Rast, and Ramsay); *The geology of Scotland* (1965; particularly the chapters by Johnson and Mercy); and the *Scottish Journal of Geology* symposium on north-east Scotland (Stewart and others, 1970).

The works in our selection are the following (some including accounts of Old Red Sandstone and Mesozoic) :

Barrow, 1893 [297]. Metamorphism in the south-eastern Highlands

Clough, Maufe, and Bailey, 1909 [372]. Cauldron-subsidence of Glen Coe

Bailey, 1910 [377]. Recumbent folds

Bailey, 1913 [397]. Loch Awe syncline

Bailey, 1916 [412]. Ben Nevis and Glen Coe

Bailey, 1922 [434]. Structure of the south-west Highlands

Anderson (E. M.), 1923 [440]. Schiehallion

Read, 1923 [447]. Banff, Huntly, and Turriff

Read, 1923 [448]. Arnage district, Aberdeenshire

Tilley, 1924 [457]. Comrie area, Perthshire

Bailey and Lee, 1925 [459], Mull etc

Read, 1925 [464]. Golspie, Sutherland

Tilley, 1925 [465]. Metamorphic zones in the southern Highlands

Read, 1927 [481]. Igneous and metamorphic history, Deeside

Read, 1931 [510]. Central Sutherland

Bailey, 1934 [527]. West Highland tectonics : Loch Leven to Glen Roy

Read, 1934 [531]. Unst, Shetlands

Read, 1935 and later editions [541]. Grampian Highlands

Read, 1935 [542]. Igneous rocks, Haddo House district, Aberdeenshire

Phemister, 1936 and later editions [552]. Northern Highlands

Read, 1936 [553]. Unst, Shetlands

Anderson (J. G. C.), 1937 [556]. Etive granite, Glen Coe

Phillips, 1937 [559]. Fabric study of Moines

Anderson (E. M.), 1942 [571]. The dynamics of faulting

Phillips, 1945 [578]. Microfabric of Moines

Kennedy, 1946 [581]. Great Glen fault

Anderson (J. G. C.), 1947 [585]. Highland Border

Anderson (E. M.), 1948 [593]. Lineation and petrofabrics

Kennedy, 1948 [596]. Thermal structure

Kennedy, 1949 [604]. Metamorphic zones in the Moines

Phillips, 1951 [611]. Life-history of the Moines

Read, 1952 [614]. Metamorphism and migmatization, Ythan valley, Aberdeenshire

Read and Farquhar, 1952 [615]. Arnage district, Aberdeenshire
Anderson (J. G. C.), 1955 [625]. Between Glen Roy and
Monadliath mountains, Inverness
Read and Farquhar, 1956 [631]. Buchan anticline
Sutton and Watson, 1956 [632]. Dalradian, Banffshire
Johnson and Stewart (editors), 1963 [648]. Caledonides
Craig (editor), 1965 [651]. Scotland

56 The Midland Valley of Scotland

We have selected some of the more notable of the works on the
stratigraphy and economic (as distinct from the volcanic) geology
of this region. Our knowledge is summarised in the *Regional geology*
handbook (2nd edition, 1948) and both summarised and enlarged
in the chapters by Waterston and Francis in *The Geology of
Scotland,* 1965 [651].
Geikie, 1902 [346]. East Fife. Most of the field-work had been
done long before.
Clough and others, 1910 [319]. East Lothian
Macgregor (M.), 1925 [462]. Contribution to the Glasgow district
Carruthers and others, 1927 [476]. Oil-shales of the Lothians
Macgregor (M), 1930 [502]. Scottish Carboniferous stratigraphy
Macgregor (M.) and MacGregor (A. G.), 2nd edition, 1948 [550].
Midland Valley of Scotland
Eyles and others, 1949 [601]. Central Ayrshire
Kennedy, 1958 [637]. Tectonic evolution
Craig (editor), 1965 [651]. Scotland

57 The Southern Uplands of Scotland

The present state of our knowledge of the succession, structure,
and palaeogeography of this area has been admirably summarised
and enriched by (1) Pringle in the *Regional geology* handbook,
2nd edition, 1948 [540], and (2) Walton in several chapters in the
British Caledonides edited by Johnson and Stewart, 1963 [648] and
in the *Geology of Scotland* edited by Craig, 1965 [651]. Going back
in time we find in the great memoir of the Geological Survey, *The
Silurian rocks of Scotland,* 1899 [327] the Ordovician, established
by Lapworth in 1879, being still included in the 'Silurian' by the
official Survey even so late as the end of the century. In this work

the history of previous research is fully reviewed, a review which, so far as the realisation of truth only is concerned, is almost entirely one of the work of Charles Lapworth, and particularly of his two papers on the Moffat area, 1878 [236] and the Girvan area, 1882 [250]. The first of these papers is one of those few treatises that not only establish the essential geological features of a region, sweeping the board clean of all previous and largely erroneous accounts, but also at the same time fix the truth by such an abundant display of detailed facts (in a key area) that they stand complete, once and for all. In the history of the recognition of those significant geological facts which enable research to move onwards into new fields, this 'Moffat paper' is of paramount importance. It demonstrated that graptolites, in their different kinds, could be used in a refined application of William Smith's principle of 'strata identified by organised fossils', and that the several individual strata so discriminated could be recognised and mapped, and thus be made to reveal the structure of the region. It was found that what had hitherto been regarded as an enormously thick mass of strata was in fact a series of repetitions by folding.

> 'The caution requisite in proving and reproving every important point, stratigraphical and palaeontological, in a region so excessively disturbed has necessitated the accumulation of a mass of confirmatory and supplementary evidence, sufficient to place wholly beyond cavil all the data upon which our conclusions are founded' (Lapworth, 1878).

Lapworth's work in general has been reviewed by Watts (1939) and by Boulton and others (1951). Among the papers and monographs published in the present century the following are in our selection : W. A. Deer on the Cairnsmore of Carsphairn igneous complex, 1935 [534] and A. Williams on the structural history, 1959 [641] and the Caradocian, 1962 [646] of the Girvan district.

58 Further researches on the Tertiary volcanoes and their products in Scotland

The first period in the investigation of the volcanic rocks and volcanic history of Scotland has been taken as our theme 43. It ended with the publication of Geikie's *Ancient volcanoes* in 1897.

Geikie himself inaugurated the second period when he secured the official services of Alfred Harker for the survey of the Tertiary volcanic islands of the Inner Hebrides, starting with the most important, Skye, in 1895. After nine years the famous *Tertiary igneous rocks of Skye* [355] was published, followed in 1908 by the description of the *Small isles of Inverness-shire* [369]. Mention may be made here to Harker's posthumous sketch-book, *The West Highlands and Hebrides* (1941) with introduction by Sir Albert Seward.

The third period, begun under Clough before the first World War, and later continued largely under Bailey, produced the following works by the geologists of the Geological Survey—all on the Tertiary districts:

Bailey, Thomas, and others, 1924 [451]. Mull etc

Tyrrell, 1928 [491]. Arran

Richey, Thomas, and others, 1930 [505]. Ardnamurchan etc

Richey, 1932 [518]. Ring structures

Davidson, 1935 [533]. Raasay

Summaries and historical reviews are in the following: The histories of the Geological Survey by Flett (1937) and Bailey (1952); the *Regional geology* handbook by Richey and others on the *Tertiary volcanic districts*, 3rd edition, 1961 [543]; chapters in *The geology of Scotland*, edited by Craig, 1965 [651], particularly those by E. H. Francis (Carboniferous-Permian) and F. H. Stewart (Tertiary).

These second and third periods in the study of Scottish volcanology were thus concerned chiefly with the Tertiary activity.

59 History of igneous activity in the British region

Among the most important and certainly among the longest anniversary addresses of the President of the Geological Society were the two delivered by Sir Archibald Geikie in 1891 and 1892 [288]. They formed a review, in two consecutive parts, of the history of 'volcanic action during each great geological period' in the British region. These addresses were overshadowed a few years later by the publication of Geikie's massive two-volume work, *The ancient volcanoes of Great Britain* in 1897 [318], which includes the whole of the British Isles. This is perhaps the greatest single work ever to have been written by one author on a subject in general British geology, and among his many works which are great literature on great themes it stands out as Geikie's most splendid monument. On questions of description, discrimination, and nomenclature the searcher finds himself going back again and again to these mighty volumes, and always the point at issue is resolved. Of course his 'masterly account would today need to be much further amplified

in order to include the results of later researches', as Harker remarked in his presidential address to the Geological Society in
1917 [422], in which address Harker gave his 'rough summary of
igneous action in Britain', a highly significant review including
philosophical aspects of his subject. (Geikie had not discussed the
history of plutonic intrusion.) At about the same time Harker contributed important sections, in chronological order, to the British
Isles volume of the *Handbuch der regionalen Geologie* which
was later published in Britain as the *Handbook of the geology of
Great Britain,* 1929 [494] (theme *66*). A concise and well-documented
account, 'Igneous activity in the British Isles' which includes the
more recent work, is to be found in the latest edition of Hatch's
Petrology of the igneous rocks, rewritten with A. K. and M. K. Wells
in 1961 [290].

60 Petrology and igneous action

The ten-year period from 1907 to 1917 produced some notable
researches on igneous petrology and igneous action, and we have
selected the following in our annals :

Bemrose, 1907 [365]. The toadstones of Derbyshire

Flett, 1910 [381]. His petrological chapter in the second edition
of the *Edinburgh* memoir of the Geological Survey, as representative of his contributions to several of the Scottish memoirs.

Dewey and Flett, 1911 [385]. Pillow-lavas and the 'spilitic suite'

Thomas, 1911 [390]. Skomer volcanic series, Pembrokeshire

Tyrrell, 1917 [421]. Lugar sill, Ayrshire. (With further papers in
1948 and 1952.) This sill has provided an interesting study in stages
of crystallisation (Phillips, 1968).

61 Textbooks of petrology

The *Petrology of the igneous rocks* by F. H. Hatch, 1891 [290]
inaugurated a new era of textbooks on petrology. The book has
been for eighty years one of the most widely used geological textbooks and has been revised many times. In 1926 A. K. Wells joined
Hatch in 'a complete revision' and during the last twenty years his
son, M. K. Wells, has been a third co-author. 'The subject-matter
of the original book has now been completely replaced, but it is

considered expedient to retain the sub-title "Hatch and Wells" which has become familiar' (12th edition, 1961).

Harker's *Petrology for students: a guide to the study of rocks in thin slices* [312] was first published in 1895 and has gone through many editions, the latest being that of 1954. This textbook, with its concise descriptions, many clear illustrations of the 'thin slices', and leading references, has been a guide to many generations of students particularly to those working under the author in the Sedgwick Museum at Cambridge.

The same writer's *Natural history of igneous rocks,* 1909 [374], petrogenic rather than petrographic, is a splendid example of geological thought expressed in literature.

Hatch and Rastall's *Petrology of the sedimentary rocks,* 1913 [399], now in its fourth edition revised by J. T. Greensmith 1965 has for long been the companion volume to 'Hatch and Wells' on the igneous rocks.

Holmes's *Petrographical methods and calculations,* 1921 [430] The title speaks for itself.

Milner's *Sedimentary petrography* [437]. This is another textbook which having begun in a small way in 1922, has always given the geologist so exactly what has been wanted that its author has risen to the occasions that have demanded new editions until the latest, the fourth edition of 1962 has become a two-volume treatise of 1358 pages. Principles, applications, methods a vast amount of factual information, all lavishly illustrated and supported and enriched by a bibliography of thousands of references, both listed and, most important, cited at every relevant point in the text.

Tyrrell's *Principles of petrology,* 1926 [474] though reprinted many times has never gone beyond the very slightly revised edition of 1929. Nevertheless it is still one of the most useful of all books on petrology.

No one has done more work on the petrography of British sedimentary rocks than P. G. H. Boswell, and his book *On the mineralogy of sedimentary rocks,* 1933 [520] is a well-known landmark.

62 Descriptive palaeontology in the twentieth century

We list without comment the descriptive monographs and papers selected for our annals.

PALAEOZOOLOGY
Crinoids. Wright, 1950 [607], Carboniferous

Brachiopods. Reed, 1917 [420], Girvan, Ayrshire

Lamellibranchs. Arkell, 1929-37 [493], Corallian

Gasteropods. Harmer, 1914-24 [404], Pliocene

Cephalopods. Buckman, 1909-30 [371], Type ammonites; Spath, 1923-43 [449], Gault ammonites; Arkell, 1935-48 [532], Corallian ammonites; Arkell, 1951-8 [608], Bathonian ammonites; Currie, 1954 [620], Scottish Carboniferous gonialites

Trilobites. Reed, 1903-35 [351], Girvan, Ayrshire; Lake, 1906-46 [362], Cambrian; Whittard, 1955-67 [630], Ordovician, Shelve, Shropshire

Insects. Bolton, 1921, [428], Coal Measures; Bolton, 1930 [499], S. Wales coalfield

Fishes. Woodward, 1902-12 [349], Chalk; Traquair, 1903 [352], Carboniferous, Edinburgh (Distribution); Woodward, 1916-19 [418], Wealden and Purbeck

PALAEOBOTANY

Seward, 1900-4 [337], Jurassic

Arber, 1904 [354], Culm Measures

Arber, 1912 [391], Coal Measures, Forest of Dean

Arber, 1916 [411], Coal Measures, S. Staffordshire

Kidston and Lang, 1917-21 [419], Old Red Sandstone, Rhynie, Aberdeenshire

Kidston, 1922-6 [436], Carboniferous

Heard, 1927 [479], Old Red Sandstone, Brecon

Heard, 1939 [566], Old Red Sandstone, S. Wales

It may be mentioned here that in 1957 appeared the first number of *Palaeontology,* published by the newly-founded Palaeontological Association. The technical papers are beautifully presented in text and picture.

The history of British palaeontology has been briefly reviewed by Cox (1956) and by the present writer (1967).

63 *Textbooks of palaeontology*

For fifty years, from the eighteen-nineties to the nineteen-forties, the student's one standard textbook, in its successive editions, was *Palaeontology: invertebrate* by Henry Woods, 1893 [302]. This, on 'systematic' lines, is still much used. In 1920 it was joined by the *Introduction to palaeontology* by A. Morley Davies, [425] which includes sections on vertebrates and plants and concentrates on selected types. The latest edition, 1961 has been revised by Sir

James Stubblefield. In 1920 was also published H. L. Hawkins's stimulating review of principles and successive faunas, his *Invertebrate palaeontology* which is a commentary rather than a practical guide. In 1923 we had H. H. Swinnerton's *Outlines of palaeontology* [450], making the third of the most widely-used general textbooks. It includes the vertebrates and emphasises evolutionary history. Its second edition appeared in 1947.

Three other important works of the late twenties and early thirties were E. Neaverson's *Stratigraphical palaeontology* [489], first published in 1928 and greatly enlarged in its second edition of 1955; A. C. Seward's splendid, well-documented, *Plant life through the ages*, 1931 [511], and Morley Davies's *Tertiary faunas* (a textbook in effect), in two-volumes, 1934-5 [528].

Several less formal but thoroughly scientific and authoritative works appearing recently are extremely useful. For instance J. F. Kirkaldy's Fossils (1967) is in treatment an attractive elementary textbook. Especially valuable handbooks from the British Museum (Natural History) are *The succession of life through geological time* (1949); the three works edited by Errol White, Keeper of Palaeontology, illustrating *British fossils: Palaeozoic* (1964), *Mesozoic* (1962), and *Caenozoic* (1959); and the 1967 edition of *The early history of palaeontology*.

64 Evolutional palaeontology

The lack of a thoroughly satisfactory harmony between the fossil record and the theory of evolution has presented a most intractable problem ever since Darwin so clearly exposed it in the tenth and eleventh chapters of *The origin of species*, 1859 [181]: those 'On the imperfection of the geological record' and 'On the geological succession of organic beings.'

It seems rather odd that the publication of Darwin's book did not at once cause a rush among palaeontologists to find more evidence as to the manner in which new fossil species appeared in the rocks. One might have thought that attempts in evolutional palaeontology would have been very much to the fore in the eighteen-sixties, but it was not till nearly the end of the century that deliberate studies began to be made in Britain.

Fossils constitute the only evidence we have as to the course of evolution in the past. It is curious that detailed evidence of the phylogenetic relationships of genera and species is much less common and certain than might have been expected, with the exception of that for the rise and fall of groups, including some now extinct, which is shown so abundantly in the fossil record.

The fact that at least one group, the graptolites, revealed a connected story of gradual change came to be realised towards the end of the nineteenth century, and this has become much more evident in the present century (theme 69). In 1899 A. W. Rowe published the results of his intensive collection and critical examination of some thousands of specimens of echinoids belonging to the genus *Micraster* from the Chalk of southern England [328], and he was able to demonstrate a very gradual evolution in form and structural details. Other well-known studies in this connection are R. G. Carruthers's on the Carboniferous coral genus *Zephrentis*, 1910 [378], H. Woods's on the lamellibranch *Inoceramus* in the Cretaceous period, 1912 [396], and A. E. Trueman's on the Lower Liassic members of the lamellibranch family, the Ostreidae, 1922 [439]. The evidence that fossils provide regarding evolution has been reviewed by the present writer (1959).

65 *Four textbooks*

We include in our annals a number of textbooks which, while casting an eye the world over, deal particularly with the geology of Britain or are profuse in their British examples. Most of them come under the headings of particular subjects; we here refer to four others.

James Geikie's *Structural and field geology*, first published in 1905 [358], has no doubt always been the most popular of his several textbooks and like all the others is beautifully illustrated, the landscape and close-up views being nearly all from the innumerable striking examples that Scotland provides. It has a wider scope than its title might imply. The sixth edition, largely revised by two other Scottish geologists, R. Campbell and R. M. Craig, was published in 1952.

'Lake and Rastall' [382] was perhaps the most widely-used textbook of geology from the time of its first publication in 1910 until some time after its last appearance in a fifth edition (with further reprints) revised by Rastall in 1941. The first part, by Rastall, is general, but inevitably cites and illustrates interesting British examples; the second part, by Lake, gives a concise but very clear account of the stratigraphy of the British Isles. The sections across the more important regions in the structure of Britain form a neat series.

Alfred Harker's *Metamorphism: a study in the transformations of rock-masses* [515], first published in 1932 and revised by the author for a second edition, 1939 shortly before his death (a third edition appeared in 1950) is one of those deeply-penetrating works for which the author is so greatly admired. British examples are very largely drawn on.

The principles of physical geology by Arthur Holmes [575] was at once acclaimed as the most desirable book on the subject when it first appeared in 1944, and the acclaim was renewed when twenty years later the vastly expanded 'new and fully revised edition' appeared. World-wide in scope, British geology occupies less proportionate space here than in the other three works.

66 The geology of England and Wales, Great Britain, and the British Isles

We refer here to works on the geology of Britain, published during the present century, some of which must be on every geologist's bookshelf for perusal and reference.

The first of these is the Jubilee Volume of the Geologists' Association, *Geology in the field,* edited by Monckton and Herries in four parts, 1909-10 [376]. This contains thirty-four authoritative regional essays which, although not formally arranged, did in fact provide the most recent views at that time of the geology of most parts of England and Wales. Twenty years later, in 1929, came the *Handbook of the geology of Great Britain* [494], a new and extended edition of a volume forming part of a projected world-wide *Handbuch der regionalen Geologie,* which had been published in Germany in 1917. This work contains systematically arranged chapters written by the leading authorities of the day, the whole edited by J. W. Evans and C. J. Stubblefield. It at once became the standard account of the geology of Great Britain; a separate account of the geology of Ireland had already appeared.

During the years 1935, 1936, and 1937 the Geological Survey issued the first editions of their *British regional geology* series of 'handbooks' [535-41, 543, 544, 547-50, 552, 555, 557, 560]. There are eighteen in all, seventeen published in those years, the remaining one appearing in 1948 [599] when the second editions of the others were issued. All continue to appear in new editions and it is quite safe to say that taken together they constitute an indispensable, and an 'official', account of the geology of Great Britain, region by region.

The *Outline of historical geology* by A. K. Wells was first published in 1938 [565] and the fifth revised edition appeared in 1966. 'Historical geology' is here exemplified exclusively from the British Isles (and what better region could there be for such an exemplification?), so that it falls naturally under the present heading. Ever since it first appeared it has been, in its successive editions with the collaboration of J. F. Kirkaldy, a standard textbook.

Finally, we have recently had a trio of works neatly disposed in their subject matter so as to cover the three main aspects of the geology of the whole region of the British Isles: stratigraphy, structure, and history. Each provides a most usefully detailed conspectus of one of these aspects: Dorothy H. Rayner's *Stratigraphy*, 1967 [653], J. G. C. Anderson and T. R. Owen's *Structure*, 1968 [655], and G. M. Bennison and A. E. Wright's *Geological history*, 1969 [657].

67 *The geological and physiographical history of the British region*

As the geological history of a region is what geology is ultimately endeavouring to decipher, writings which attempt to depict from time to time the view so far attained are obviously important. As regards the British region as a whole Edward Hull's *Contributions to the physical history of the British Isles* (1882), was the first systematic treatise on the successive geographical phases of the British region, but it was a comparatively slight sketch. A. J. Jukes-Browne's well-known *Building of the British Isles*, 1888 [270], mentioned in theme *38*, was a very much more detailed assessment of the evidence and was for long the standard work in its several editions, the best-known perhaps being the third, 1911. A fourth appeared in 1922 after the author's death. The subject was again put on a new footing by the publication of L. J. Wills's *Physiographical evolution of Britain*, 1929 [497], followed by the same author's invaluable *Palaeogeographical atlas of the British Isles and adjacent parts of Europe*, 1951 [613]. Recently we have had the Geological history of the British Isles by George M. Bennison and Alan E. Wright, 1969 [657] referred to in theme *66*.

Of studies of particular regions, periods, or aspects of the matter, the following are among the more important: Jones's 'Evolution of a geosyncline [Lower Palaeozoic], 1938 [562], and his 'Geological evolution of Wales and the adjacent regions', 1955 [628]; Wills's *Palaeogeography of the Midlands*, 1948 [598]; George's 'Lower Carboniferous palaeogeography of the British Isles', 1958 [636] and his 'Tectonics and palaeogeography in England', 1962-3 [645]; and Allen's 'Wealden environment: Anglo-Paris basin', 1959 [638]. Ziegler (1970) gives a recent review in his 'Geosynclinal development of the British Isles during the Silurian period'.

68 Geological maps of Great Britain, the United Kingdom, and the British Isles

The geological maps of England and Wales published in the second half of the nineteenth century were mentioned in theme 49. We refer here to the general maps of the present century, on the scales of 25 miles to the inch and 10 miles to the inch.

The small-scale map of the British Isles is extremely well-known, its size and bright colouring forming a picturesquely striking wall-map. It was first issued by the Geological Survey under Teall's directorship in 1906 [364], and successive editions were published in 1912, 1939, 1957, and 1969. In 1966 was issued a map of an entirely new kind, the 'Tectonic map of Great Britain and Northern Ireland' prepared by F. W. Dunning under Stubblefield's director-ship, [652]. The map shows the distribution of the structural elements, with the character and form of the main individual folds and faults, and by sections on the sheet, the structure at depth. It is a triumph in that such a vast amount of significant, though necessarily to some extent conjectural, information (much of it the result of recent work), is presented so clearly and compactly.

In 1948 was published by the Geological Survey, with McLintock as director, the '10-mile' map of Great Britain in two sheets, which at once became the standard map of the whole country [597]. A second edition in 1957 was reprinted with minor emendations in 1964.

69 Graptolites, the succession of graptolite faunas, and the evolution of the graptolites

Lapworth's epoch-making paper on 'The Moffat series' in the Quarterly Journal of the Geological Society, 1878 [240] gave a detailed account of the succession of the 'Glenkiln', the 'Hartfell' and the 'Birkhill' and the 'Gala' (Llandovery) series, with an enumeration of the graptolites characteristic of the successive beds. The graptolite zones of this large part of the Lower Palaeozoic were determined. The zones of the rocks below (Arenig) were very largely added by Elles from the succession which she interpreted in the Skiddaw Slates in 1898 [320], and those in the rocks above by Elles again in 1900 [333], from the Wenlock of the Welsh Borderland, and by Wood in the same year from the Ludlow of the same region [338]. Then the Misses Elles and Wood, under the direction of Professor Lapworth, began the preparation of the *British graptolites* monograph [340], which was produced in parts during the next twenty years. All these works described and illustrated the different species and recorded in great detail and with meticulous exactness their occurrence in the successive beds. On the basis of all this factual information the purely biological consideration naturally arose : could a continuous evolution of form within the group be traced? Preliminary suggestions were made by Nicholson and Marr in the *Geological Magazine* for 1895, but it was, fittingly, Dr. Elles who gave the fundamental exposition in 1922 [435]. Since then detailed studies of the graptolites have been made by Dr. Bulman, and in 1958 he summarised the position as regards our view of their evolution in an address to the Palaeontological Association [635].

'It was Lapworth's "Geological distribution of the Rhabdophora" [241] that finally established the stratigraphical value of the graptolites, but any general discussion of the succession of graptolite faunas may take as its starting-point "The graptolite faunas of the British Isles", published by Miss Elles in 1922. In this, she expresses the evolutionary and philosophical conclusions reached after more than twenty years' work in collaboration with Lapworth and Miss Wood in the preparation of the famous *Monograph* (1901-19). The evolutionary history of the entire group was analysed in terms of certain general trends—stipe reduction, change in direction of growth, and various trends in thecal elaboration—and stress was laid on their stratigraphical significance (Bulman).

The zones established by Lapworth, Elles and Wood remain the present standard, which is a proof of the great thoroughness with which the work was done. This is all the more remarkable in that it was practically new work from the start, though some work had been done, and was concurrently being done in Sweden. In Lapworth's work the sequence of the graptolites, when once established in a local condensed succession of the black shales, became the key to the structure of the whole of the Southern Uplands; in the Welsh Borderland the simpler structure enabled the graptolite sequence to be read through extended successions over a wide area.

The two papers by Dr. Gertrude Elles and Miss Wood (Dame Ethel Shakespear) are contiguous in more senses than one in the Geological Society's *Quarterly Journal,* and taken together and combined with the *Monograph* they constitute one of the earliest, and probably for all time the most notable, contributions by women geologists to British geology. A very full history of the growth of knowledge about graptolites constitutes a separate volume of the *Monograph.* This monograph is unique in that it describes and figures virtually every species of a distinctive, varied, and abundant group. Any newly-collected British graptolite will still be found to have in this monograph (completed more than fifty years ago) a form with which it may either be positively identified or very closely compared.

Detailed research has recently been in progress on the morphological and stratigraphical palaeontology of the graptolites, particularly those from Silurian rocks (e.g. Rickards, 1967).

70 *The North-west Pennines*

The Caroniferous Limestone, with the Whin Sill enclosed within it, forms the conspicuous westward-facing escarpment in the north-west of England, and at intervals along its foot inliers of Lower Palaeozoic, with perhaps older rocks, occur. The Ingleborough district in particular forms one of those special highlights which are conspicuous in a general view of the geology of Britain.

In the first place this district provides the best example we have of unconformity, revealed by field-mapping as in an idealised plan-diagram, visible in distant views across the valleys, and strikingly displayed in some quarry-sections. We have already noted Playfair's discovery of two exposures (1802, theme *13*). There had been a Geological Survey memoir in 1890, but McKenny Hughes's series of papers published by the Yorkshire Geological Society, 1901-8 [345] give a more informal account, with sketches and historical allusions and much detail, particularly about the Lower Palaeozoic rocks, and with discussion of the physical features and such well-known local peculiarities as the perched-blocks and limestone-weathering effects. This is perhaps the most interesting of the geological publications, as distinct from the archaeological ones, by this genial and revered Cambridge professor. The district has been concisely yet fully and attractively described and illustrated, with full references, in a much more recent article of the same society's *Proceedings* (Dunham and others, 1953).

K

The other chief inlier is the one between the town of Appleby and that eminence on the Carboniferous plateau, Cross Fell. We have mentioned Buckland's early description in 1817 (theme *24*). Here we note the titles of two papers in our selection of the more modern era: those by Nicholson, Marr, and Harker, 1891 [293], and Shotton, 1935 [545]. The more recent paper by Dean (1959, on the Caradocian) should be mentioned, and that in preparation by Burgess and others on the Silurian.

71 *The Jurassic and Cretaceous rocks of Britain*

JURASSIC

The history of the growth of our knowledge of the Jurassic rocks of England and Wales during the last hundred years largely centres successively on the work of particular geologists: Blake, Buckman, Fox-Strangways, Woodward, Richardson, Lang, and Arkell.

When J. F. Blake published in 1875 the first of his extensive stratigraphical and palaeontological papers in the *Quarterly Journal* of the Geological Society on the Jurassic formations, very little had been done during the preceding forty years. This first paper on 'The Kimmeridge Clay of England' [217], was followed by 'The Corallian rocks of England' (with W. H. Hudleston) in 1877, [229] and 'The Portland rocks of England' in 1880 [243]. Meanwhile he and R. Tate had published their book on *The Yorkshire Lias,* 1876 [223].

S. S. Buckman's life-work was on the zonal stratigraphy of the Inferior Oolite of the Gloucester-Somerset-Dorset outcrop. By pushing too far his thesis that ammonites could be used as infallible markers of very small geological time-divisions, he did in fact demonstrate the limits to which they could properly be so used; and he used them properly in tracing the non-sequences and oversteps within the formation. His work, our selections of which [275, 298, 311, 316, 339] are from 1889 to 1901, has been meticulously reviewed by Morley Davies in a memorial address to the Geologists' Association (1930).

We come next to 'the most comprehensive memoir ever published by the Geological Survey' (Flett, 1937), of which the principal authors were Fox-Strangways, 1892 [294] and H. B. Woodward, 1893-5 [303].

'Although such gigantic compilations as *The Jurassic rocks of Britain* (in 5 volumes, although Scotland and Ireland are not included) have long remained invaluable sources of facts, they labour under a fundamental disability, for by striving after an illusory completeness they become more and more encumbered with detail as new discoveries are made—a negation of progress. Unless the endless description of sections and listing of fossils be regarded as a worthy aim in itself, it will be agreed that the discovery of new exposures, the filling in of gaps, should lead to the elimination of detail and enable generalizations to be made' (Arkell, 1933).

This fork includes historical summaries.

Linsdall Richardson's work, which has been briefly reviewed in the Geological Society's obituary notice in 1967, was chiefly on the Rhaetic, Lias, and Inferior Oolite of Gloucester and Somerset. His researches were largely published in the Cotteswold Naturalists' Field Club *Proceedings,* and in 1904 appeared his attractive *Handbook to the geology of Cheltenham and neighbourhood* [356]. His most important papers were published in the *Quarterly Journal* of the Geological Society, 1907-18 [367, 417] on districts stretching from the Rissingtons in the north to Crewkerne in the south.

W. D. Lang's papers on the Lias of the Dorset-Devon coast (and parts inland) in the region of Charmouth, Lyme Regis, and Pinhay are well-known for their detailed stratigraphy and palaeontology of that famous strip. The series in the Geological Society's *Quarterly Journal,* 1923-8 [445, 471, 487] was in collaboration with L. F. Spath and others, while those in the Geologists' Association *Proceedings,* 1914-32 [405], with their large, large-scale maps are among the most notable publications of that society. The Geological Society's obituary notice on Lang also appeared in 1967; Richardson and Lang were long-lived contemporaries.

During the nineteen-thirties, forties, and fifties one name completely dominated the Jurassic geology and palaeontology of Britain : that of William Joscelyn Arkell, with a wonderful succession of detailed papers, monographs, memoirs, and books.

'From 1929 to 1948 much of his time was occupied in the compilation of two large monographs of the Palaeontographical Society, on the Corallian lamellibranchs [493] and ammonites [532], the illustrations of which are an example of his outstanding technical skill in photography. It was, however, in 1933 that Arkell established himself as a leading authority on Jurassic stratigraphy and palaeontology by the publication of his *Jurassic system in Great Britain* [519], a monumental work of nearly 700 pages with copious plates, a remarkable achievement for a man not yet thirty years of age. It contained not only a detailed account of the stratigraphy but embodied a historical review of the principles of classification and correlation enunciated by d'Orbigny, Oppel, Buckman and others, together with a demonstration of the relationship between sedimentary facies, troughs of deposition, and axes of folding' (Douglas, 1959).

The historical review to which Douglas refers in the above quotation is not confined to Britain. It forms a chapter of 37 pages and is undoubtedly one of the best essays to be found anywhere on the history of geology.

In 1947 Arkell's *Geology of Oxford* appeared, and in the same year the Geological Survey memoir on the *Country around Weymouth, Swanage, Corfe and Lulworth* [586] of which he was the chief author. These books deal with the Cretaceous and later formations as well as with the Jurassic. Arkell's *Jurassic geology of the World* is an exhaustive treatise on the extra-British regions.

CRETACEOUS

Companion volumes to those on the Jurassic are the three on *The Cretaceous rocks of Britain* by A. J. Jukes-Browne [335], with contributions by W. Hill, 1900-4. Notwithstanding the title the work is confined to England. There are detailed summaries in each of the three volumes of the history of the recognition and nomenclature of the subdivisions of the system.

Papers on the Cretaceous in our selection include important ones in the *Quarterly Journal* of the Geological Society on the Lower Cretaceous of coastal regions in eastern England : two by Lamplugh on the Speeton Clay in Yorkshire, 1889 [278], 1896 [315] and one, forty years later, by Swinnerton on Lincolnshire 1935 [546]. This region is covered by Wilson in the *East Yorkshire and Lincolnshire Regional geology* handbook, [599]. Between 1900 and 1908 a series of papers on 'The zones of the White Chalk of the English coast' by A. W. Rowe appeared in the Geologists' Association *Proceedings* [336]. These are beautifully illustrated with photographs on which the horizons are precisely indicated.

MESOZOIC OF SCOTLAND

Between 1873 and 1878 Judd wrote three papers in the *Quarterly Journal* of the Geological Society on 'The Secondary rocks of Scotland' [212], but the second of these was concerned more with the Tertiary volcanics than with these 'secondary' (Mesozoic) rocks (see theme 43). In 1932 was published Lee and Pringle's 'synopsis of the Mesozoic rocks of Scotland' [516] in the Glasgow society's *Transactions*.

72 *Various districts of England*

Among the important miscellaneous works on particular districts of England published during the first four decades of the twentieth century and which have not been described in our other themes, we list here the following seven from our annals.

G. W. Lamplugh, 1903 [350]. Isle of Man

Lamplugh was allotted the task of surveying the Isle of Man when 'a new field was opened' by the Geological Survey in that area.

'He began in 1892 and finished the field survey in 1897. The one-inch map was published in 1898 and the memoir in 1903. In the petrographical work Lamplugh had the able support of Watts, and together they produced a map and memoir which furnish a complete account of the geology of the island' (Flett, 1937). 'The Manx Slates of the Isle of Man were thought by Lamplugh to be equivalent to the lower and largely concealed part of the Skiddaw Slates. No fossils of diagnostic value have been found, and by some [most] authors, including Lamplugh, the rocks are referred provisionally to the Cambrian [including the Tremadocian] system' (Eastwood, *Regional geology* handbook on *Northern England,* 3rd edition, 1963 [536]).

E. A. N. Arber, 1911 [384]. The Coast scenery of North Devon

This, in the present writer's opinion, is the best work ever to have been written on coastal geology and geomorphology in Britain. Sound principles are expounded and convincingly exemplified in text and beautiful photographs. It is the perfect example of a 'semi-popular' book being at the same time a scientific 'classic'.

'It has occurred to me that a general account of the geology of the scenery of this coast-line, expressed in as simple a language as possible, might interest those who visit North Devon and North Cornwall. I desire that this volume should be readable and of use to visitors who are interested in geology, in a greater or less degree.'

J. S. Flett, 1912 [392]. Lizard peninsula, Cornwall

There is a good account of the Lizard peninsula, in the *Regional geology* handbook on *South-west England,* 3rd edition, 1969 [535], a district which has produced so much petrological description of

the serpentines and associated igneous rocks, and so much speculation as to structure and ages.

P. G. H. Boswell, 1916 [413]. Eocene of the north-east part of the
London basin

This doctorate's thesis, published in the *Quarterly Journal* of the
Geological Society, is an outstanding essay on the structural and
palaeogeographical implications of stratigraphical petrology.

V. C. Illing, 1916 [414]. The Middle Cambrian rocks of the
Nuneaton inlier

Here there is a detailed succession of horizons 'teeming with
beautiful fossils' of the Paradoxidian fauna, chiefly the small
trilobite *Agnostus*. The quotation is from Nicholas's second paper
[408] (read at the same meeting of the Geological Society) in which
a similar but not so copious a fauna is described from St. Tudwal's
peninsula, Caernarvonshire.

L. D. Stamp, 1921 [433]. The conception and illustrations of cycles
of sedimentation in the Eocene of the Anglo-Franco-Belgian basin;
further discussed in his *Introduction to stratigraphy* (1923)

P. G. H. Boswell, 1923 [441]. Petrography of the Cretaceous and
Tertiary outliers of the west of England

P. F. Kendall and H. E. Wroot, 1924 [456]. *Geology of Yorkshire:
an illustration of the evolution of northern England*

One of the greatest books ever to have been written on one of
the larger regions of Britain. It is in two volumes, of over a thousand
pages. Much attention is given to the history of human endeavour
among the rocks. Volume 2 has two parts; the first is entitled
'Yorkshire from a railway carriage window', a method of travel
which lends itself excellently to a transection of geological views.
The second part has a similarly engaging title: 'Specimen days in
Yorkshire'.

R. G. Carruthers and others, 1932 [513]. The Cheviot Hills

A Geological Survey memoir. This interesting igneous complex
is well summarised in the *Regional geology* handbook *Northern
England* by Eastwood, 3rd edition 1963 [536].

T. N. George, 1940 [568]. Structure of Gower

One of the author's important papers on the Carboniferous Limestone areas of South Wales. 'By mapping the surface distribution
of the Avonian zones it has been possible to analyse in detail the
Armorican structures in Gower'.

W. W. Watts, 1947 [592]. Charnwood Forest

Published in the year of the author's death, it describes the
'ancient rocks' of that curiously revealed inlier of Pre-Cambrian
composition and Triassic landscape.

73 The Lake District

Clifton Ward's *Geology of the northern part of the English Lake District*, 1876 [226] was the first published result of the Geological Survey's invasion of the most northerly parts of Britain. A number of papers had been written on the Lake District since Sedgwick's time (theme *30*) by Harkness, Nicholson, and others and these are listed as an appendix to Ward's memoir. This list is one of those invaluable productions of William Whitaker whose bibliographical researches into the literature of British regions were not the least of his great contributions to geology. Ward's memoir is especially remarkable as being the first publication of the Geological Survey to contain descriptions and figures of the microscopic characters of rocks, these being supplemented by the more particular studies published at just about the same time in the *Quarterly Journal* of the Geological Society [220, 221] (theme *48*).

In 1888 appeared the paper on the Stockdale Shales by J. E. Marr and H. A. Nicholson [272], who were among the first to apply in England Lapworth's discovery and demonstration in Scotland of the use of graptolites in stratigraphy. In 1891 we had the paper on the Shap Granite and its metamorphism by Harker and Marr [289] with its supplement in 1893, and in 1894 and 1895 Harker's papers on the Carrock Fell gabbro and granophyre [304]; two studies from opposite ends of the district of great significance regarding the natural history of igneous action. Rastall's account of the Skiddaw Granite and its metamorphism, 1910 [383] is to be added to these.

Meanwhile Gertrude Elles had made her study of the graptolites of the Skiddaw Slates, 1898 [320] (theme *69*).

In 1906 Marr delivered to the Geological Society his presidential address on 'The influence of the geological structure of English Lakeland upon its present features' [363], and in 1916 appeared his well known work *The geology of the Lake District* [416].

Papers on the Lake District continued to be published and the next period of research has been reviewed by Hollingworth in the Geologists' Association *Proceedings* for 1954. Since then, two publications by G. H. Mitchell (who had four important papers in the Geological Society's *Quarterly Journal* between 1929 and 1940) are to be mentioned: his 'Geological history of the Lake District' in the Yorkshire Geological Society's *Proceedings* (1956), and his Geological Association *Guide* (1970). The district is well covered in the *Regional geology* handbook on *Northern England* by Eastwood, 3rd edition, 1963 [536].

74 Pleistocene geology

We list first of all one publication of each of those two complementary topics : raised beaches and submerged forests :

Prestwich, 1892 [296]. The raised beaches of the south of England.

Reid, 1913 [401]. *Submerged forests*

Secondly we list in chronological order the selected publications dealing with glaciation :

Harker, 1901 [343]. Ice-erosion in the Cuillin Hills, Skye

Kendall, 1902 [348]. A system of glacier lakes in the Cleveland Hills

Davis, 1909 [373]. Glacial erosion in North Wales

Smith, 1912 [395]. The glaciation of the Black Combe district (Cumberland)

Wright, 1914 [407]. *The Quaternary ice age*

Trechmann, 1915 [409]. The Scandinavian Drift of the Durham coast

Lamplugh, 1920 [427]. Pleistocene glaciation of England. (Presidential address to the Geological Society)

Wills, 1924 [458]. The development of the Severn valley in the neighbourhood of Ironbridge and Bridgnorth, Shropshire

Trotter, 1929 [496]. The glaciation of eastern Edenside, the Alston block, and the Carlisle plain

Hollingworth, 1931 [509]. Glaciation of western Edenside and the Solway basin

Dewey, 1932 [514]. The palaeolithic deposits of the lower Thames valley

Charlesworth, 1955 [627]. The later-glacial history of the Highlands and islands of Scotland

Charlesworth, 1957 [634]. *The Quaternary era*. One of the great works by an author who has made the subject his own, grasped the whole matter with a masterly hand, and who presents it in all its detail and history as a complete work of art, science and literature. Its scope is world-wide.

Thirdly, there are the works dealing with the 'Crags' and later deposits of East Anglia. Until recently these 'crags' were all classified as 'Pliocene', but now only the lowest of these, the Coralline Crag is so placed. The works dealing with the 'Pliocene' in general of

East Anglia are thus largely works on the Pleistocene as now
defined. Our selection comprises the following :

Wood, 1880-2 [247]. This paper, in two instalments in the
Quarterly Journal of the Geological Society is on what the author
calls the 'newer Pliocene' which in fact exactly corresponds to the
present 'Pleistocene' as it excludes the Coralline Crag below and
includes the glacial and later deposits above.

Reid, 1890 [282]. The Geological Survey memoir, *The Pliocene
deposits of Britain*

Harmer, 1898-1900 [322]. Two papers on the 'Pliocene deposits
of the east of England' : the first deals with the Coralline Crag and
thus with the Pliocene in the now restricted sense, the second with
the Pleistocene.

Boswell, 1927-9 [475]. Several Geological Survey memoirs on the
southern part of East Anglia

Chatwin, 1961 [557]. The *Regional geology* handbook on *East
Anglia and adjoining areas*, 4th edition 1961

75 Devonian and Old Red Sandstone

Our selection from the works published during this period contains titles dealing with some of the main areas of outcrop. In
South Devon the chief name is that of W. A. E. Ussher of the
Geological Survey, who was author or part-author of eleven district
memoirs in explanation of the corresponding map-sheets during the
period 1902 to 1913. We have selected one of the earliest and most
notable of these, that of the 'country around Torquay', [353]. In
1890 had appeared, in the *Quarterly Journal* of the Geological
Society, Ussher's paper on the 'Devonian rocks of South Devon'
[283].

Wickam King's important paper on the Downtonian and
Dittonian of Britain and north-western Europe, 1934, [530] was the
result of his research on the then wilderness of the Shropshire-
Worcester-Hereford Old Red Sandstone. King himself wished to
place both of these series in the Silurian. The Dittonian is
universally placed in the Old Red Sandstone, but there is no
absolute agreement at present as to whether the Downtonian should
be included in the Silurian or the Old Red Sandstone, though
opinion inclines to placing it in the latter, the Devonian system,
with the Ludlow Bone Bed at its base.

The Old Red Sandstone of Caithness, particularly its chief
constituent the Caithness Flagstone series, is one of the most

distinctive of all British stratigraphical formations, with its rhythmic lithological sequences and above all its fossil fishes. The Geological Survey memoir, 1914 [403] is by Crampton and Carruthers.

Finally there are the two papers by Allan on the Old Red Sandstone of the Highland Border, 1928, [485], 1940 [567].

76 *Lower Carboniferous*

The exact study of the stratigraphy and correlation of the Lower Carboniferous rocks could only be undertaken by the application of zonal methods. This was made possible by Arthur Vaughan's investigation of practically the whole succession in the excellently exposed section in the Avon gorge west of Bristol. His paper published in 1905 let loose a flood of research which was one of the most notable features in the history of British geology for the next thirty years. The fossil groups found to be most useful were in the first place, the corals, and secondly, the brachiopods. The whole matter was reviewed up to 1926 by S. H. Reynolds in his address to the British Association; and for the modern position we turn to such summaries as those provided in the recent general treatises on the geology of Britain by, for instance, Wells and Kirkaldy [565], Rayner [653], and Bennison and Wright [657].

We here list the works on this subject which are in our selection :
Vaughan, 1905 [360]. 'Palaeontological succession' in the Bristol area
  Sibly, 1908 [370]. Derbyshire and North Staffordshire
  Dixon and Vaughan, 1911 [386]. Gower
  Garwood, 1912 [393]. North-west England
  Vaughan, 1915 [410]. Correlation between the British and Belgian successions
  Silby and Dixey, 1918 [423]. South-eastern margin of the South Wales coalfield
  Hudson, 1924 [455]. Yoredale series (rhythmic succession)
  Reynolds, 1926 [472]. General review
  George, 1927 [478]. Northern margin of the South Wales coalfield
  Lewis, 1930 [501]. Isle of Man
  Garwood, 1931 [508]. Cumberland and Roxburghshire
  George, 1933 [525]. Western part of the Vale of Glamorgan
  Dunham and Stubblefield, 1945 [576]. Greenhow, Yorkshire
  Hudson and Cotton, 1945 [577]. Alport boring, Derbyshire
  George, 1958 [636]. Palaeogeography

Regional geology handbooks, particularly those on the *Bristol and Gloucester district* [547], *South Wales* [560], the *Pennines* [555], and *Northern England* [536]

The Lower Carboniferous rocks in the north of England have been reviewed in detail by Rayner (1953).

77 Upper Carboniferous

MILLSTONE GRIT SERIES (NAMURIAN)

Hind and Howe, 1901 [344]. The beds between the 'Millstone Grit' and the 'Limestone-massif' at Pendle Hill, Lancashire, and their equivalents elsewhere. One of the first attempts to grapple with the problem of the stratigraphical relationships of the beds between the limestones and the massive grits in various parts of northern England. The matter has been discussed by the present writer (1949-51).

Gilligan, 1920 [426]. Petrography of the Millstone Grit of York-shire. The well-known paper with its frequently reproduced phy-siographical map which discusses, largely from petrographical data, the conditions under which the Millstone Grit was deposited.

Bisat, 1924 [452]. Goniatites of the north of England and their zones. One of those papers that completely revolutionised the study of stratigraphy by applying zonal methods. After the ammonites in the Jurassic, the graptolites in the Ordovician-Silurian, the corals in the Lower Carboniferous—now the goniatites of the Namurian and contiguous beds. At last the puzzle of their correlation could be solved.

Hudson and Cotton, 1943 [572] and 1945 [577]. The detailed suc-cession and structure at Alport Dale, Derbyshire, revealed in borings for oil

Dunham and Stubblefield, 1945 [576]. The Millstone Grit part of the description of the Greenhow mining area, Yorkshire

COAL MEASURES AND COALFIELDS—GENERAL

Trueman and Davies, 1927 [483], Dix and Trueman, 1931 [507], Trueman 1933 [526], and Trueman, 1947 [591]. The first two of these papers carried the Namurian zones up into the Coal Measures, but here by means of the non-marine lamellibranchs. Another momentous straightening-out of correlation was thus effected. A broad view of this correlation is given in Trueman's 1933 paper, and a still broader view of the 'stratigraphical problems in the coal-fields of Great Britain' is presented in his address to the Geological

Society in 1947. Finally, this great authority edited and helped to write *The coalfields of Great Britain,* 1954 [624]. Still more recently, as the result of Survey work, coalfield correlation has been reviewed by Woodland and others (1957).

COAL MEASURES—SOUTH WALES

Strahan and others, 1899-1921 [329]. The great series of thirteen Geological Survey memoirs constituting the *Geology of the South Wales coalfield.* Later editions of some of these were published.

The following three papers on several aspects : Dix, 1934 [529], on the sequence of floras; Trotter, 1947 [590], on the structure of the Pontardawe-Ammanford area; and Trotter, 1949 [606], on the devolatilisation of the coal seams.

COAL MEASURES—MIDLANDS AND NORTHERN ENGLAND

Gibson, 1901 [341]. Upper Coal Measures

Gibson, 1905 [359]. Geological Survey memoir on the *North Staffordshire coalfields*

Hickling, 1927 [480]. Sections of strata of the Coal Measures of Lancashire

Dix, 1931 [506]. Flora of the upper portion of the Coal Measures of North Staffordshire

STAFFORDSHIRE

Fearnsides, 1933 [524]. Correlation of structures in the coalfields of the Midland province

MIDLAND PROVINCE

Edwards and Stubblefield, 1948 [594]. Marine bands, marker horizons, and sedimentary cycles in the Middle Coal Measures of Nottinghamshire and Derbyshire

COAL MEASURES—SCOTLAND

Weir and Leitch, 1936 [554]. The zonal distribution of the non-marine lamellibranchs in the Coal Measures of Scotland

Leitch, 1941 [570]. Upper Carboniferous rocks of Arran

MILLSTONE GRIT AND COAL MEASURES

The chief *Regional geology* handbook dealing with these formations is that on *the Pennines and adjacent areas,* 3rd edition, 1954 [555].

78 Pre-Cambrian and Lower Palaeozoic rocks of North Wales

CAERNARVONSHIRE AND NORTHERN MERIONETH—THE CAMBRIAN AND ORDOVICIAN

Fearnsides, 1905 [357]. The volcanic rocks (Carodician) of the mountain of Arenig Fawr and the Cambrian (Lingula Flags—Tremadoc series) and Arenig sedimentary rocks immediately to the west

Fearnsides, 1910 [380]. The Cambrian (Lingula Flags—Tremadoc series) and lowest Ordovician of the Tremadoc district

Nicholas, 1915-16 [408]. The Cambrian (Harlech Grits—Tremadoc series) and Ordovician (chiefly Arenig series) of St. Tudwal's peninsula : mainly a detailed account of the Cambrian and in the second paper descriptions of the Middle Cambrian trilobites.

Fearnsides and Morris, 1926 [469]. Stratigraphy and structure of the Cambrian slate-belt of Nantlle on the north-west side of the Snowdon syncline

Fearnsides and Davies, 1944 [573]. The Cambrian (Lingula Flags—Tremadoc series) and Ordovician of the 'Deudraeth' area, east of the Tremadoc area described in 1910.

Mention should here be made of Fearnsides's suggestive account of North Wales in *Geology in the field,* 1910 [376].

LLEYN PENINSULA—THE PRE-CAMBRIAN AND ORDOVICIAN

Matley, 1913 [400]. Bardsey Island (Pre-Cambrian)

Matley, 1928 [488]. Pre-Cambrian of south-western Lleyn

Matley and Heard, 1930 [503]. Volcanic and sedimentary rocks, of the middle part of the Ordovician, of Garn Bodfean (near Nevin) and its immediate surroundings

Matley, 1932 [517]. Volcanic and sedimentary rocks of Arenig—Llanvirn age of Mynydd Rhiw towards the south-west tip of the peninsula

Matley and Smith, 1936 [551]. The 'Sarn granite', S. W. Lleyn

Matley, 1938 [563]. The Ordovician rocks of the country round Pwllheli, Llanbedrog, and Madryn

ANGLESEY

Greenly, 1919 [424]. The famous two-volume Geological Survey memoir

Greenly, 1923 [442], 1930 [500]. Papers detailing further researches on the Mona Complex (Pre-Cambrian)

CADER IDRIS AND DOLGELLEY

Cox and Wells, 1921 [429]. The Lingula Flags—Llanvirn sedimentary and associated igneous rocks of the Arthog-Dolgelley area

Cox, 1925 [460]. Cader Idris. The igneous rocks and the Lingula Flags—Bala succession

LLANGOLLEN DISTRICT

Wills and Smith, 1922 [438]. Tectonics of the Lower Palaeozoic rocks

CORRIS AND LLANOWCHLLYN—ORDOVICIAN

Pugh, 1923 [446], 1928 [490], 1929 [495]. Three papers constituting a major piece of research among the hitherto unexplored rocks of upper Ordovician and lower Silurian age, lying in a belt of country to the south-east of the ridge of middle Ordovician volcanic rocks, stretching from Cader Idris to the Arans. R. M. Jehu's paper (1926) completes the survey between Corris and the coast at Towyn.

CENTRAL CAERNARVONSHIRE

Williams, 1927 [484]. Snowdon. The Cambrian and (particularly) the Ordovician igneous and associated sedimentary rocks of Snowdon and its ridges

Williams and Bulman, 1931 [512]. The 'Dolwyddelan syncline'. The middle part of the Ordovician including the 'Snowdon volcanic suite'

DENBIGHSHIRE—SLUMP STRUCTURES IN THE SILURIAN

Jones, 1937 [558], 1940 [569]. The two papers in which the hypothesis was developed that submarine sliding or slumping, contemporaneous with the deposition of the sediments, was the cause of many of the internal crumples and minor folds of rock-formations; illustrated by a description and discussion of such structures in the Ludlovian (1937) and the Salopian (1940) rocks.

THE HARLECH DOME

Matley and Wilson, 1946 [583]. Cambrian (Harlech Grits to Ffestiniog Flags) of the Harlech Dome, between the Barmouth estuary and the vale of Ffestiniog

'The following account is dedicated to the memory of Professor

Charles Lapworth, who initiated and inspired the study and mapping of the district now described, who carried out a large amount of the field work himself, and who established the stratigraphical succession and the geological structure of the Harlech Dome in the early years of the century, many years before his death in 1920.'

NORTH WALES—THE SILURIAN

Boswell, 1949 [600]. The Middle Silurian rocks of North Wales, chiefly Denbighshire. Boswell did not accept Jones's slumping hypothesis.

NORTH WALES—GENERAL

North, (1949). This is a very useful introduction to general geology, based on a description and discussion of a limited region. It contains a section on the history of investigation.

George and Smith, 1961 [538]. The *Regional geology* handbook, 3rd edition

79 Ordovician and Silurian rocks of Central Wales

The history of the research into the Ordovician and Silurian rocks which constitute Central Wales, and the present state of our knowledge, has been reviewed by Bassett (1969(a), as part of a wider survey) and the present writer (1969, concentrating on Cardiganshire).

The following are our selected works, grouped according to area.

THE WESTERN PART AND GENERAL

Jones, 1909 [375]. Plynlimon, Pont-erwyd, and Devil's Bridge. Detailed stratigraphy and structure, with the graptolite zones of the Valentian (Llandovery series) as there developed. The first paper in a new era of investigation of Central Wales, though the paper by Herbert Lapworth on the Rhayader district (1900), should not be forgotten.

Jones, 1912 [394]. A review and exposition of the geological structure of the whole region

Jones and Pugh, 1916 [415]. Machynlleth area. A continuation and application of the research previously carried out to the south (Jones, 1909)

Jones, 1921 [431]. The Valentian series. A wide-ranging review of the British rocks of this age, with a somewhat particular reference to Central Wales

The Rhayader—Pumpsaint area

Davies, 1926 [468]. Between the mountain-top of Drygarn and the village of Abergwesyn to the south.

Davies, 1928 [486]. Between Rhayader and Abergwesyn.

Davies, 1933 [523]. Between Abergwesyn and Pumpsaint.

THE LLANDEILO—LLANDOVERY AREA

Jones and Andrew, 1925 [461]. Three consecutive papers on the Llandovery rocks of (1) the Llandovery district, southern area (Jones); (2) the Garth district (Andrew); and (3) the relations between the two districts (Jones and Andrew)

Jones, 1949 [602]. Llandovery district, northern area

Williams, 1953 [619]. Llandeilo district

THE BUILTH—LLANDRINDOD AREA

Straw, 1937 [561]. The higher Ludlovian rocks of the Builth district

Jones and Pugh, 1946 [580]. The complex intrusion of Welfield, near Builth

Jones, 1947 [589]. Silurian rocks west and south of the Carneddau range

Jones and Pugh, 1948 [595]. A multi-layered dolerite laecolite near Llandrindod, and other dolerite masses in the Builth—Llandrindod inlier

Jones and Pugh, 1949 [603]. An early Ordovician shore-line near Builth

It will be seen that the name of Owen Thomas Jones dominates the whole of this subject. His work has been described in the obituary notice by his frequent associate, William John Pugh (1967).

80 The Permo-Trias

Sherlock carefully examined the evidence as to the time-classification and correlation of the beds usually grouped into two separate systems, Permian and Trias: in detail in the north-east of England, [388] and in a general review over the British Isles, 1926-8 [473]. The lithological divisions are largely a matter of facies, and correlation by fossils is impossible. Hence the calling of all these rocks the 'Permo-Triassic formations' or the grouping of them into one system, the 'Permo-Trias'. The Magnesian Limestone where it occurs is in the lower part of the whole; the sandstones, conglomerates, and marls had long been known as the New Red Sandstone. In 1947 Sherlock published a world review, *The Permo-Triassic formations.*

Trechmann in three papers on the Magnesian Limestone of County Durham, 1913 [402], 1914 [406], 1925 [466] gave a thorough account of all aspects of its geology and palaeontology in the 'classical' area of this notable formation which, in addition to containing characteristic fossils, is lithologically unique in Britain.

Boulton's two papers 1924 [453], 1933 [521] described and discussed the beds between undoubted Upper Carboniferous and undoubted Trias in the Birmingham district, chiefly from information supplied by borings. Should they be classed with the Carboniferous or placed as Permian?

The Permo-Triassic system being the most extensive of those outcropping in the 'Central England' district we here refer to the account of them in the *Regional geology* handbook on that region by Edmunds and Oakley, 3rd edition, 1969 [549].

81 The geological time-scale

'The development of a satisfactory quantitative time-scale, in which the times of geological events are defined in the same absolute units of time as are used by the historian and astronomer is an achievement of the present century'. This is the first sentence of the section by L. R. Wager, 'The history of attempts to establish a quantitative time-scale' in the Geological Society's symposium-volume *The Phanerozoic time-scale*, edited by Harland, Smith, and Wilcock, 1964 [650]. For further remarks on this volume see theme *86*. The main stages in the formulation of the geological time-scale by radiometric methods are the three expositions by Holmes in 1911 [387], 1947 [588], and 1959 [639].

One of the most important single papers on the absolute geo-chronology of British rocks is that dealing with the metamorphic complexes of the Scottish Highlands by Giletti, Moorbath, and Lambert, 1961 [643]. These complexes are chiefly Pre-Cambrian, i.e. Pre-Phanerozoic or Cryptozoic.

82 Concealed geology

Any direct information as to the rocks below the visible surface

L

of the ground is naturally of the utmost importance, and such information, providing actual specimens from precisely located points,
can only be obtained from underground workings (particularly
for coal) or by borings and the recovery of the cores from them.
Very interesting facts were obtained by Davies and Pringle 1913
[398], from borings for water at 'Calvert station' in Buckinghamshire. A succession (with non-sequences) in the Jurassic was found
to rest directly (and of course unconformably) on Tremadocian
shales of the Cambrian.

> 'The geological basis of the recent search for oil in Great
> Britain by the D'Arcy Exploration Company was described
> by Lees and Cox in 1937. That paper showed how investiga-
> tions, carried out through a number of years, had resulted in
> an accumulation of evidence sufficient to justify a new ex-
> ploration campaign on a more extensive and systematic scale
> than in any previous attempt.'

> 'The search for oil in Great Britain [in this new campaign]
> has been carried out in five separate geological provinces—
> the Mesozoic of southern England, the Carboniferous of the
> eastern Midlands, the Triassic and Carboniferous of
> Lancashire, the Permian of North Yorkshire, and the Calci-
> ferous Sandstone Series of the Midland Valley of Scotland.'

These quotations are from the important paper by Lees and Taitt,
1945 [582]. In 1960 was published the Geological Society's memoir
by Falcon and Kent, on *Geological results of petroleum exploration
in Britain, 1945-1957* [642].

> 'G. W. Lamplugh and F. L. Kitchin in 1911 added to know-
> ledge of the underground structure of Kent in a memoir deal-
> ing with the concealed rocks of the Kent coal explorations.
> This knowledge has been amplified by further publications on
> the coalfield area, notably by Lamplugh, Kitchin, and J.
> Pringle in 1923, by H. G. Dines, R. Crookall and C. J. Stubble-
> field in 1933 and by Sir A. E. Trueman and the last-named
> in 1946.'

This quotation is from the 4th edition, 1965 of the *Regional geology*
handbook on *The Wealden* district by Edmunds, [537], in which
these concealed Triassic and Jurassic strata are summarised. The
work by Lamplugh, Kitchin, and Pringle mentioned in the quotation is in our selection [444].

In 1954 W. B. R. King, in his presidential address to the
Geological Society on 'The geological history of the English

Channel' [621] gave geological maps of that area and of the neighbouring coastal areas of England and France. There is very little recent sedimentary cover so that here the outcrops are mostly concealed merely by the water. Since then this work has been greatly supplemented and in 1970 the Geological Survey published an important report by Curry and others. It should be mentioned that in 1966 the 'Geological Survey of Great Britain', together with overseas geological surveys, became incorporated as the 'Institute of Geological Sciences'.

83 Three miscellaneous subjects: seismology, mineral resources, 'way-up' of strata

The two kinds of geological phenomena which are liable to be catastrophically destructive to human life and property are volcanic action and earthquakes. The evidence of volcanic action in Britain in past geological times we have noted under several heads. General works on seismology, the science of earthquakes, are chiefly concerned with regions outside Britain, but for Britain we have Davison's *History of British earthquakes,* 1924 [454].

We are hardly concerned with 'economic geology' in this review, but as one representative work we list Sherlock's *Rock-salt and brine,* 1921 [432] from the *Mineral resources* series of memoirs of the Geological Survey.

In reading a local rock-succession it is obviously of the first importance to know that we are not reading it backwards, that is, to be sure of the 'way-up' of the strata. Bailey has gone into this matter in his review, in the *Geological Magazine* for 1949, of Shrock's *Sequence in layered rocks* (1948), and as one of the most significant papers we list his own in the same magazine, 1930 [498].

84 The Lower Palaeozoic rocks of Shropshire

The Pre-Cambrian and Lower Palaeozoic region of South Shropshire has for long been a favourite ground for geologists in search of pleasure and profit in their science. The fourth visit of the Geologists' Association was made in 1925 and for that visit Professor Watts wrote an account [467], with full references, which brought up to date the account he had written for a previous excursion in 1894 (theme 50). A fifth excursion was held in 1952 for which a full account of the area [616] incorporating the recent work was written by the leader, Professor Whittard. He himself had carried out detailed researches on the stratigraphy and palaeonto-logy of the Valentian (Llandoverian) rocks, 1928-32 [492], and later produced a monograph for the Palaeontographical Society, 1956-67 [630] on the Ordovician trilobites of the Shelve area.

In 1909 E. S. Cobbold made his first 'report' (continued till 1916) to the British Association on excavations carried out in the Cambrian rocks of Comley, and between 1910 and 1920 published, in the *Quarterly Journal* of the Geological Society, descriptions of the fossils, nearly all of which were from the Lower and Middle Cambrian. It is curious that the smallest and most poorly exposed outcrops of Cambrian rocks in southern Britain—here, in South Shropshire, and at Nuneaton—are the most prolific in fossils. In 1927 appeared his paper [477] in which the stratigraphy and structure of this area were gathered together with maps and sections. In 1933 a detailed account of the Lower and Middle Cambrian rocks of Rushton near the Wrekin was published by Cobbold and R. W. Pocock in the *Philosophical Transactions of the Royal Society* [522].

Meanwhile in 1927, just fifty years after the appearance of Callaway's paper (theme 50), Stubblefield and Bulman published their paper on the Shineton Shales [482], with full lists and des-criptions of the numerous small but sometimes well-preserved fossils, which, in addition to their purely palaeontological interest, enabled stratigraphical horizons to be discriminated.

In the Geological Survey memoir on the *Shrewsbury district* (Pocock and others, 1938) the history of research among the Cambrian rocks is traced. Finally in 1968 and 1969 respectively appeared the Geological Survey maps (1/25,000) of the Church Stretton and Craven Arms areas, and the corresponding booklets written by J. E. Wright and B. A. Hains.

The *Regional geology* handbook on *The Welsh Borderland* by Pocock and Whitehead, 2nd edition, 1948 [539] gives a summary.

85 *The Northamptonshire ironstone field*

The Northamptonshire ironstone industry—the working of the bedded ironstone that occurs in the Northampton Sand formation of the Inferior Oolite series—has necessitated an investigation of the geology and has afforded exceptional opportunities for this investiga-tion because of the extent of the workings. The area is of special interest, apart from the geology of the ironstone, because it has provided remarkable examples of 'superficial structures' resulting from movements due to flowage of clay under load. The distribution and origin of these structures was discussed by Hollingworth, Taylor, and Kellaway in 1944 [574]. These are the three names associated with the investigation of all aspects of the geology of the ironstone field and neighbouring region during the next twenty years. In 1946 an account of the Kettering district was written by Hollingworth and Taylor (of the Geological Survey) for the field meeting of the Geologists' Association [579], and three Survey memoirs by Taylor (one in collaboration with Hollingworth) on various aspects were published in 1949 [605], 1951 [609], and 1963 [649]. In 1953 Kellaway and Taylor discussed in detail the physiographical evolution during the Pleistocene of part of the East Midlands, with a further consideration of the superficial structures [617].

86 Some recent collaborative volumes

It has recently become the custom to hold symposium meetings taking a particular subject which calls for a collaboration comprising contributions by specialists : critical compilations, essays, expositions of new knowledge—with discussions. The results of such a local gathering may later be given to the world in the form of a symposium volume. Such a volume may be made by inviting contributions of this kind without necessarily holding a formal meeting, and one of the most notable features in the publication of geological literature during the decade of the nineteen-sixties was the appearance of a number of these. In the following notices the quotations are taken from the books themselves.

The British Caledonides, edited by M. R. W. Johnson and F. H. Stewart, 1963 [648]. Symposium meeting held at Edinburgh University, December 1961. There are ten contributors. The 'Caledonides' are, in the first place and as originally defined, the (inferred) mountains produced by the 'Caledonian' mountain-building revolution during Lower Palaeozoic and Devonian times, but the term is used in a comprehensive sense for the whole orogenic belt.

> 'Our knowledge of the British Caledonides is the result of the work of many famous geologists over the past hundred years or more. In recent years spectacular advances have been made following the development of new techniques of structural and sedimentary analysis. The various lines followed by modern

workers in different parts of the Caledonides—structure, sedimentation, mineralogy and petrology, geochemistry, isotope
radiometric geology, etc—are now really beginning to come
together in a form which should lead to a major synthesis. It
is therefore useful at this point to have a kind of Caledonian
stocktaking'.

The Phanerozoic time-scale, edited by W. B. Harland, A. Gilbert
Smith, and B. Wilcock, 1964 [650]. Papers presented at a joint meeting of the Geological Society of London and the Geological Society
of Glasgow held at the University of Glasgow in February 1964.
There are twenty-two contributors.

'The *Phanerozoic time-scale* is a compilation of radiometric
data for the construction of a time-scale (in years) in terms of
Cambrian-to-Tertiary stratigraphy, and a critical re-examina-
tion of this time-scale from both the stratigraphical and radiometric viewpoints. The treatment is international in scope,
with the aim of producing a publication of world-wide significance. The volume is dedicated to Professor Arthur Holmes
as the doyen of British geologists working in the field of geo-
chronology.'

The Geology of Scotland, edited by Gordon Y. Craig, 1965 [651].
There are twelve contributors. The title speaks for itself and we
have already referred to the work in themes *54, 55, 56, 57, 58.* The
Preface contains a sketch of the progress of knowledge of Scottish
geology.

The fossil record, edited by W. B. Harland and eight others, 1967
[654]. Symposium meeting held at the University College of Swansea
in December 1965, jointly organised by the Geological Society of
London and the Palaeontological Association.

'The fossil record sets out to depict, document, and analyse
the stratigraphical range, as published, of all fossil taxa [classi-
fication-units] covering comprehensively the whole range of
plants and animals in the [world-wide] fossil record. It is a
symposium volume to which more than a hundred specialists
have contributed.'

The geology of the East Midlands, edited by P. C. Sylvester,
Bradley and T. D. Ford, 1968 [656]

'This work, by seventeen authors, provides a chronological
account of the geological history of the East Midlands over
the last 1000 million years. Each chapter sets out to survey
the present state of knowledge of a formation, to bring up to

date the interpretation of that knowledge, and to incorporate
the results of recent research work. A bibliography of about
1000 works is included.'

The Pre-Cambrian and Lower Palaeozoic rocks of Wales, edited
by Alan Wood, 1969 [659]. Symposium meeting held at the University College of Wales, Aberystwyth in January 1967, in honour of
Professor Owen Thomas Jones and Sir William John Pugh. There
are twenty-eight contributors.

'Wales is classic ground for the study of Lower Palaeozoic
rocks. The system-names, Cambrian, Ordovician and Silurian
commemorate Wales (Cambria) and the names of two ancient
Welsh tribes, the Silures and the Ordovices. In geological
circles the names of small market towns like Bala, Llandeilo
and Llandovery, a wild and lonely mountain, Arenig, even an
isolated farm house in Pembrokeshire, Llanvirn, have gained
international currency.'

Time and place in orogeny, edited by P. E. Kent, G. E. Satterthwaite, and A. M. Spencer, 1969 [658]. Papers presented at a joint
meeting of the Geological Society of London and the Yorkshire
Geological Society held at the University of Durham in January
1968.

'*Time and place in orogeny* contains 15 original papers
planned to cover problems encountered in the geological
analysis of orogenic belts.'

Appendix A

Secondary and associated literature

NOTE

The numbers refer to the Themes (Section II)
in which the works are mentioned

Adams, F. D. 1938 *The birth and development of the geological sciences*. Baltimore. *Preface, 1*

Arkell, W. J. 1956 *The Jurassic geology of the World*. Edinburgh, Oliver and Boyd. *71*
——and Tomkeieff, S. I. 1953 *English rock terms*. London, Oxford University Press. *1, 25*

Bailey, E. B. 1949 Review of R. R. Shrock's *Sequence in layered rocks*. (New York, McGraw-Hill, 1948). *Geol. Mag.*, 86 : 132-4. *83*
——1950 James Hutton, founder of modern geology. *Proc. Roy. Soc. Edinb.*, 63(B) : 357-68. *9*
——1952 *Geological Survey of Great Britain*. London, Murby. *42, 45, 54, 58, 59*
——1962 *Charles Lyell*. London, Nelson. *35*
——1967 *James Hutton: the founder of modern geology*. Amsterdam, Elsevier. *9*

Bassett, D. A. 1967 *A source-book of geological, geomorphological and soil maps for Wales and the Welsh Borders*. Cardiff, National Museum of Wales. *49*
——1969(a) Some of the major structures of early Palaeozoic age in Wales and the Welsh Borderland : an historical essay, in *The Pre-Cambrian and Lower Palaeozoic rocks of*

168

Wales, edited by Alan Wood, 67-116. Cardiff, University of Wales Press.   79

——1969(b) Wales and the geological map. *Amgueddfa (Bull. National Museum of Wales),* no. 3, 10-25.   2

Bonney, T. G. 1895 *Charles Lyell and modern geology.* London, Cassell.   35

Brander, G. 1754 A dissertation on the belemnites. *Phil. Trans. Roy. Soc.,* 48 : 803-10.   5

Bromehead, C. E. N. 1945 Geology in embryo (up to 1600 A.D.). *Proc. Geol. Ass.,* 56 : 89-134.   *I*

Brooks, M. 1970 Pre-Llandovery tectonism and the Malvern structure. *Proc. Geol. Ass.,* 81 : 249-68.   46

Burgess, I. C. and others. (In preparation) The Silurian strata of the Cross Fell area.   70

Butcher, N. E. 1962 The tectonic structure of the Malvern Hills. *Proc. Geol. Ass.,* 73 : 103-23.   46
——(c. 1968) W. G. Maton and the geological map of S.W. England. *Proc. Ussher Soc.,* 2 : 14.   16

Challinor, J. 1945 Dr Plot and Staffordshire geology. *Trans. N. Staffs. Fld Club,* 79 : 29-67.   3
——1947 From Whitehurst's *Inquiry* to Farey's *Derbyshire. Trans. N. Staffs. Fld Club,* 81 : 52-88.   4, 8, 25
——1948 The beginnings of scientific palaeontology in Britain. *Ann. Sci.,* 6 : 46-53.   20
——1949 Thomas Webster's observations on the geology of the Isle of Wight, 1811-13. *Proc. Isle of Wight Nat. Hist. Arch. Soc.,* 4 : 108-22.   26
——1949-51 North Staffordshire geology, 1811-1948. *Trans. N. Staffs. Fld Club,* 83-85 : A1-A64.   77
——1951 Jonathan Otley's 'Geology of the Lake District'. *North-western Naturalist,* 23 (for 1948): 113-26.   30
——1953-4 The early progress of British geology. *Ann. Sci.,* 9 : 124-53, 10 : 1-19, 107-48.   *Preface*
——1959 Palaeontology and evolution. in *Darwin's biological work,* edited by P. R. Bell, 50-100. Cambridge, University Press.   64

——1961-4 Some correspondence of Thomas Webster, geologist (1773-1844). *Ann Sci.*, 17 : 175-95; 18 : 147-75; 19 : 49-79, 285-97; 20 : 59-80, 143-64.   *26*

——1967 A brief review of discoveries in British palaeontology. *Welsh Geol. Quarterly*, 2 : 3-11.   *62*

——1967 *A dictionary of geology*, 3rd edition. Cardiff, University of Wales Press. (1st edition, 1961)   *Preface*

——1968 Uniformitarianism—the fundamental principle of geology. *Report of the International Geological Congress*, 23rd session, 13 : 331-43.   *35*

—1969 Geological research in Cardiganshire, 1842-1967. *Welsh Geol. Quarterly*, 4 : nos. 2 and 3, 3-37. (Earlier version, for period 1842-1949, in *Ceredigion*, 1, 1951 : 144-76)
                                                                36, 51, 79

——1970 The progress of British geology during the early part of the nineteenth century. *Ann. Sci.*, 26 : 177-234.   *Preface, 29*

Chorley, R. J. and others. 1964 *The history of the study of landforms*. Vol. 1 : London, Methuen.   *18, 35, 40, 53*

Clark, J. W. and Hughes, T. McK. 1890 *The life and letters of Adam Sedgwick*. Cambridge, University Press.   *36*

Colbert, E. H. 1962 *Dinosaurs: their discovery and their world*. London, Hutchinson.   *32*

Conybeare, W. D. 1832 *Rep. Br. Ass. Advmt. Sci.* for 1831 and 1832, 365-414.   *27*

Cox, L. R. 1942 New light on William Smith and his work. *Proc. Yorks. Geol. Soc.*, 25 : 1-99.   *19, 27*

——1948 *William Smith and the birth of stratigraphy*. Pamphlet prepared for the International Geological Congress, 18th session, London.   *27*

——1956 British palaeontology : a retrospect and survey. *Proc. Geol. Ass.*, 67 : 209-20.   *3, 62*

Curry, D. and others 1970 *Geological and shallow subsurface geophysical investigations in the western approaches to the English channel*. London, Geol. Surv.   *82*

Curwen, E. C. (editor). 1940 *The journal of Gideon Mantell*. Oxford, University Press.   *32*

Davies, A. M. 1930 The geological life-work of Sydney Savory Buckman. *Proc. Geol. Ass.*, 41 : 221-40. *71*

Davies, G. L. 1969 *The earth in decay.* London, Macdonald.
18, 35, 40, 53

Davis, A. G. 1943 The Triumvirate : a chapter in the heroic age of geology. *Proc. Croydon Nat. Hist. Sci. Soc.*, 11 : 122-46.
28
——1952 William Smith's geological atlas and the later history of the plates. *Jnl. Soc. Bibliography Nat. Hist.*, 2 : 388-95. *27*

De la Beche, H. T. 1830 *Sections and views illustrative of geological phaenomena.* London, Treuttel and Würtz. *33*
—— 1848 Anniversary address of the President. *QJGS*, 4 : (*Proceedings*) xxv-cxx. *32*

Dean, W. T. 1959 The stratigraphy of the Caradoc series in the Cross Fell inlier. *Proc. Yorks. Geol. Soc.*, 32 : 185-227. *70*

Donovan, D. T. 1966 *Stratigraphy: an introduction to principles.* London, Murby. *52*

Dott, R. H. 1969 James Hutton and the concept of a dynamic earth. in *Toward a history of geology,* edited by C. J. Schneer, 122-41. Cambridge, Mass., MIT Press. *9*

Douglas, J. A. 1959 Obituary notice of William Joscelyn Arkell. *Proc. Geol. Soc. Lond.*, session 1958-9, 141-2. *71*
——and Cox, L. R. 1949 An early list of strata by William Smith. *Geol. Mag.*, 86 : 180-8. *29*
——and Edmonds, J. M. 1950 John Phillips's geological maps of the British Isles. *Ann. Sci.*, 6 : 361-75. *49*

Dunham, K. C. and others. 1953 A guide to the geology of the district round Ingleborough. *Proc. Yorks. Geol. Soc.*, 29 : 77-115. *70*

Edwards, W. N. 1967 *The early history of palaeontology.* London, British Museum (Natural History). (Second edition of the *Guide to an exhibition*, 1931.) *3, 63*

Eyles, Joan M. 1967 William Smith: the sale of his geological collection to the British Museum. *Ann. Sci.*, 23: 177-212.
28

——1969(a) William Smith (1769-1839): a bibliography of his published writings, maps and geological sections, printed and lithographed. *Jnl Soc. Bibliography Nat. Hist.*, 5: 87-109.
27

——1969(b) William Smith: some aspects of his life and work. in *Toward a history of geology*, edited by C. J. Schneer, 142-58. Cambridge, Mass., MIT Press.
27

Eyles, V. A. 1936 Early geological maps. in *The early maps of Scotland*, 2nd edition, 77-82. Edinburgh, Royal Scottish Geographical Society.
23

——1937 John Macculloch, F.R.S., and his geological map. *Ann. Sci.*, 2: 114-29 (also 4: pl. V).
23

——1948 Louis Albert Necker, of Geneva, and his geological map of Scotland. *Trans. Edinb. Geol. Soc.*, 14: 93-127.
23

——1950 Note on the original publication of Hutton's *Theory of the Earth*, and on the subsequent forms in which it was issued. *Proc. Roy. Soc. Edinb.*, 63(B): 377-86.
9

——1955 Bibliography and the history of science. [Theories of the Earth.] *Jnl Soc. Bibliography Nat. Hist.*, 3: 63-71.
18

——1961 Sir James Hall, Baronet, 1761-1832. *Endeavour*, 20: 210-16.
21

——1969 The extent of geological knowledge in the eighteenth century. in *Toward a history of geology*, edited by C. J. Schneer, 159-83. Cambridge, Mass., MIT Press.
18

——and Eyles, Joan M. 1938 On the different issues of the first geological maps of England and Wales. *Ann. Sci.*, 3: 190-212.
27

——1951 Some geological correspondence of James Hutton. *Ann. Sci.*, 7: 316-39.
9

Falcon, N. L. 1947 Major clues in the tectonic history of the Malverns. *Geol. Mag.*, 84: 229-40.
46

Farey, J. 1813 Notes on . . . Mr. Robert Bakewell's *Introduction to geology*. *Phil. Mag.*, 42: 246-61.
22

Fitton, W. H. 1818 Geology of England. [Historical review.] *Edinburgh Review*, 29: 310-37.
27

——1832-3 Notes on the history of English geology. *Lond. and*

Edinb. Phil. Mag. Jnl Sci., 1 : 147-60, 268-75, 442-50; 2 :
37-57. *4, 29*
——1839 Lyell's *Elements of geology. Edinburgh Review,* 69 : 406-
66. *35*

Flett, J. S. 1937 *The first hundred years of the Geological Survey
of Great Britain.* London, HMSO. *42, 45, 54, 58, 71, 72*

Ford, T. D. 1960 White Watson (1760-1835) and his geological
sections. *Proc. Geol. Ass.,* 71 : 346-63. *25*
——1967 The first detailed geological sections across England, by
John Farey, 1806-8. *Mercian Geologist,* 2 : 41-9. *19*

Geikie, A. 1875 *Life of Sir Roderick Murchison.* London, Murby.
36, 45
——1871 *The Scottish school of geology.* Inaugural lecture,
Edinburgh, reprinted in *Geological sketches at home and
abroad,* London, Macmillan. 1882. *9*
——1895 *Memoir of Sir Andrew Crombie Ramsay.* London, Macmillan. *34, 45*
——1905(a) *The Founders of geology,* 2nd edition. London, Macmillan. (1st edition, 1897.) *Preface, 1, 9, 21*
——1905(b) Hugh Miller : his work and influence. in *Landscape in
history and other essays.* London, Macmillan. *45*
——1906 Lamarck and Playfair. *Geol. Mag.,* 43 : 145-53, 193-202.
9
——1918 *Memoir of John Michell.* Cambridge, University Press. *4*
——1924 *A long life's work; an autobiography.* London, Macmillan.
45, 54

Gillispie, C. C. 1951 *Genesis and geology.* Cambridge, Mass.,
Harvard University Press. (Reprinted Harper Torchbooks,
1959) *18, 31, 35*

Gordon, Elizabeth O. 1894 *The Life and correspondence of William
Buckland, D.D., F.R.S.* London, Murray. *32*

Greenly, E. 1938 *A hand through time: memories.* London, Murby.
54

Gunther, R. T. 1930 *Early science in Oxford,* vols. 6, 7 : *Life and
work of Robert Hooke.* Oxford, University Press. *3*

Hains, B. A. 1969 *The geology of the Craven Arms area.* London, Geol. Surv.   *84*

Harker, A. 1941 *The West Highlands and the Hebrides.* With an introduction by Sir Albert Seward. Cambridge, University Press.   *58*

Hicks, H. 1881 The classification of the Eozoic and Lower Palaeozoic rocks of the British Isles. *Pop. Sci. Rev.*, new series, 5 : 289-308.   *51*

Hollingworth, S. E. 1954 The geology of the Lake District—a review. *Proc. Geol. Ass.*, 65 : 385-402.   *73*

Hull, E. 1882 *Contributions to the physical history of the British Isles.* London, Stanford.   *67*

Jackson, J. F. 1932 An outline of the history of geological research in the Isle of Wight. *Proc. Isle of Wight Nat. Hist. Arch. Soc.*, 2 : 211-20.   *26*

Jackson, J. W. 1945 Martin Lister and Yorkshire geology and conchology. *Naturalist* for 1945, 1-11.   *3*

Jehu, R. M. 1926 The geology of the district around Towyn and Abergynolwyn, Merioneth. *QJGS*, 82 : 465-89.   *78*

Jones, O. T. 1921 The Valentian series. *QJGS*, 77 : 144-74.   *36*

Jones, R. K. and others. 1969 An Upper Llandovery limestone overlying Hollybush Sandstone (Cambrian) in Hollybush quarry, Malvern Hills. *Geol. Mag.*, 106 : 457-69 (issued in 1970).   *46*

Judd, J. W. 1880 On the Oligocene strata of the Hampshire basin. *QJGS*, 36 : 137-77.   *26*
——1897 William Smith's manuscript maps. *Geol. Mag.*, 34 : 439-47.   *19*
——1898 The earliest engraved geological maps of England and Wales. *Geol. Mag.*, 35 : 97-103.   *27*
——(editor). 1911 *The student's Lyell* (Introduction). London, Murray.   *35*

Keynes, G. 1960 A bibliography of Dr Robert Hooke. Oxford, Clarendon Press. 3

Kirkaldy, J. F. 1967 *Fossils*. London, Blandford Press. 63

Kuenen, P. H. 1958 Experiments in geology. *Trans. Geol. Soc. Glasg.*, 23 : 1-28. 21

Lapworth, H. 1900 The Silurian sequence of Rhayader. *QJGS*, 56 : 67-137. 79

Lees, G. M. and Cox, P. T. 1937 The geological basis of the present search for oil in Great Britain. *QJGS*, 93 : 156-94. 82

Loewinson-Lessing, F. Y. 1954 A historical survey of petrology. Translated from the Russian by S. I. Tomkeieff. Edinburgh, Oliver and Boyd. 48

Lyell, Katherine M. (editor) 1881 *Life, letters, and journals of Sir Charles Lyell, Bart*. London, Murray. 35

Mantell, G. A. 1833 *The geology of the south-east of England*. London, Longman. 32, 34

Marr, J. E. 1929 *The deposition of the sedimentary rocks*. Cambridge, University Press. 44

Mather, K. F. (editor) 1967 *Source book in geology, 1900-1950*. Cambridge, Mass., Harvard University Press. 67
——and Mason, S. L. 1939 *A source book in geology*. New York, Hafner Publishing. 4, 21

Mitchell, G. H. 1956 The geological history of the Lake District. *Proc. Yorks. Geol. Soc.*, 30 : 407-63. 73
——1970 *The geology of the Lake District*. London, Geologists' Association. 73

Muir-Wood, Helen M. 1951 The Brachiopoda of Martin's *Petrificata Derbiensia*. *Ann. Mag. Nat. Hist.*, series 12, 4 : 97-118. 20

Newbiggin, M. I. and Flett, J. S. 1917 *James Geikie: the man and the geologist.* Edinburgh, Oliver and Boyd.   *40*

Nicholson, H. A. and Marr, J. E. 1895 Notes on the phylogeny of the graptolites. *Geol. Mag.,* 42 : 529-39.   *69*

North, F. J. 1928 *Geological maps.* Cardiff, National Museum of Wales.   *27, 49*
——1931 From Giraldus Cambrensis to the geological map. *Trans. Cardiff Nat. Soc.,* 64 : 20-97.   *2, 4, 22*
——1932 From the geological map to the geological survey. *Trans. Cardiff Nat. Soc.,* 65 : 42-115.   *22*
——1933 Dean Conybeare, geologist. *Trans. Cardiff Nat. Soc.,* 66 : 15-68.   *22*
——1934 Further chapters in the history of geology in South Wales. *Trans. Cardiff Nat. Soc.,* 67 : 31-103.   *22*
——1943 Centenary of the glacial theory. *Proc. Geol. Ass.,* 54 : 1-28.   *40*
——1956 W. D. Conybeare. *Proc. Bristol Nat. Soc.,* 29 : 133-46. *47*
——1965 *Sir Charles Lyell.* London, Barker.   *35*
——, Campbell, B. and Scott, R. 1949 Snowdonia : the national park of Wales. London, Collins. 1-156.   *78*

Oakley, K. P. and Muir-wood, Helen M. 1949 *The succession of life through geological time.* London, British Museum (Natural History).   *63*

Owen, R. 1846 *A history of British fossil mammals and birds.* London, Van Voorst.   *32*

Phillips, J. 1844 *Memoirs of William Smith.* London, Murray.
    27, 29

Phillips, J. 1968 The crystallisation of the teschenite from the Lugar sill, Ayrshire. *Geol. Mag.,* 105 : 23-34.   *60*

Phipps, C. B. and Reeve, F. A. E. 1967 Stratigraphy and geological history of the Malvern, Abberley and Ledbury hills. *Geol. Jnl,* 5 : 339-68.   *46*
——1969 Structural geology of the Malvern, Abberley and Ledbury hills. *QJGS,* 125 : 1-37.   *46*

Pidgeon, E. 1830 *The fossil remains of the animal kingdom*. London, Whittaker and Treacher. *32*

Platt, J. 1764 An attempt to account for the origin and formation of . . . the belemnite. *Phil. Trans. Roy. Soc.*, 54 : 38-52. *5*

Playfair, J. 1805 Biographical account of the late Dr. James Hutton. *Trans. Roy. Soc. Edinb.*, 5 : (Reprinted in *The works of John Playfair*, Edinburgh, 1822.) *12*

Pocock, R. W. and Whitehead, T. H. 1938 *The geology of the Shrewsbury district*. London, Geol. Surv. *84*

Prestwich, Lady [G. A.]. 1899 *Life and letters of Sir Joseph Prestwich*. (With a summary of his scientific work, by Archibald Geikie.) Edinburgh, Blackwood. *42*

Raw, F. 1952 Structure and origin of the Malvern Hills. *Proc. Geol. Ass.*, 63 : 227-39. *46*

Rayner, Dorothy H. 1953 The Lower Carboniferous rocks in the north of England. *Proc. Yorks. Geol. Soc.*, 28 : 231-315. *76*

Rickards, R. B. 1967 The Wenlock and Ludlow succession in the Howgill Fells. *QJGS*, 123, 215-51. *69*

Salter, J. W. 1857 (In a communication by R. I. Murchison.) *Rep. Brit. Ass. Advmt. Sci.* for 1857 (1858), 82-4. *54*

Sandford, K. S. 1967 Obituary notice of Linsdall Richardson. *Proc. Geol. Soc. Lond.*, session 1966-7, 226-8. *71*

Sedgwick, A. 1831 Anniversary address of the President. *Proc. Geol. Soc. Lond.*, 1 : 270-316. *19*

Sheppard, T. 1917 William Smith : his maps and memoirs. *Proc. Yorks. Geol. Soc.*, 19 : 75-253. (Reprinted separately, Hull, 1920.) *34*

Sherlock, R. L. 1947 *The Permo-Triassic formations*. London, Hutchinson. *80*

M

Shrock, R. R. 1948 *Sequence in layered rocks*. New York, Mc-
    Graw-Hill.   *83*

Smiles, S. 1878 *Robert Dick, baker, of Thurso: geologist and
    botanist*. London, Murray.   *45*

Smith, S. 1942 Seventeenth century observations on rocks [etc.].
    Proc. Bristol Nat. Soc., series 4, 9 : 406-26.   *3*
——1945 Owen's *Observations*. *Proc. Bristol. Nat. Soc.*, 27 : 93-
    103.   *5*

Stamp, L. D. 1923 *An introduction to stratigraphy*. London, Murby.
                                                                 72

Stewart, F. H. and others. 1970 The 'younger' basic igneous com-
    plexes of north-east Scotland and their metamorphic
    envelope : a symposium. *Scottish Jnl Geology*, 6 : 1-132.   *55*

Stubblefield, C. J. 1951 The goniatites named in Martin's *Petrificata
    Derbiensia*. *Ann. Mag. Nat. Hist.*, series 12, 4 : 119-24. *20*

Swinton, W. E. 1965 *Fossil amphibians and reptiles*. 4th edition.
    London, British Museum (Natural History).   *32*
——1967 *The dinosaurs*. London, British Museum (Natural
    History).   *32*

Thomas, H. D. 1967 Obituary notice of William Dickson Lang.
    Proc. Geol. Soc. Lond., session 1965-66, 202-3.   *71*

Thomas, H. H. and Jones, O. T. 1912 On the Pre-Cambrian and
    Cambrian rocks of Brawdy, Haycastle, and Brimaston
    (Pembrokeshire). *QJGS*, 68 : 374-401.   *51*

Tomkeieff, S. I. 1950 James Hutton and the philosophy of geology.
    Proc. Roy. Soc. Edinb., 63(B) : 387-400.   *9*
——1962 Unconformity : an historical study. *Proc. Geol. Ass.*, 73 :
    383-417.   *4, 13*

Tresize, G. 1969 The case of the elusive *Chirotherium*. *Amateur
    Geologist*, 4 : 13-18.   *50*

Trueman, A. E. 1923 Some theoretical aspects of correlation. *Proc.
    Geol. Ass.*, 34 : 193-206.   *52*

Turner, F. J. and Verhoogen, J 1960 *Igneous and metamorphic petrology*. 2nd edition. New York, McGraw-Hill.   55

Tyrrell, G. W. 1948 A boring through the Lugar sill. *Trans. Geol. Soc. Glasg.*, 21 : 157-202.   60
——1952 A second boring through the Lugar sill. *Trans. Edinb. Geol. Soc.*, 15 : 374-92.   60

Walford, E. A. 1902 On some gaps in the Lias. *QJGS*, 58 : 267-78.   52

Warner, R. 1801 *History of Bath*. Bath.   29
——1811 *New guide through Bath*. Bath.   29

White, E. (editor). 1959 *British Caenozoic fossils*. London, British Museum (Natural History).   63
——1962 *British Mesozoic fossils*. London, British Museum (Natural History). (2nd edition, 1964.)   63
——1964 *British Palaeozoic fossils*. London, British Museum (Natural History).   63

White, G. W. 1970 Announcement of glaciation in Scotland : William Buckland (1784-1856). *Jnl Glaciol.*, 9 : 143-5.   40

White, H. J. O. 1921 *A short account of the geology of the Isle of Wight*. London, Geol. Surv.   26

Wilcockson, W. H. 1947 The geological work of Henry Clifton Sorby. *Proc. Yorks. Geol. Soc.*, 27 : 1-22.   48

Wilson, G. 1946 The relationship of slaty cleavage and kindred structures to tectonics. *Proc. Geol. Ass.*, 57 : 263-302.   41

Woodland, A. W. and others. 1957 Classification of the Coal Measures. *Bull. Geol. Surv.*, 13 : 6-13.   77

Woodward, H. B. 1908 *The history of the Geological Society of London*. London, Longmans.   27, 32
——1911 *History of geology*. London, Watts.   *Preface, 22*

Wright, J. E. 1968 *The geology of the Church Stretton area*. London, Geol. Surv.   84

Zittel, K. A. von. 1901 *History of geology and palaeontology*. Translation and abridgement by Maria M. Ogilvie-Gordon of *Geschichte der Geologie und Paläontologie*, Munich, 1899. London, Scott.   *Preface, 32*

———1932 *Textbook of palaeontology*. Vol. 2 : *Vertebrates; fishes to birds*. Translated from the German, edited by C. R. Eastman, revised by A. S. Woodward. London, Macmillan.

32

Ziegler, A. M. 1970 Geosynclinal development of the British Isles during the Silurian period. *Jnl Geol.*, 78 : 445-79.   67

Appendix B

Index of Authors, with some biographical detail

NOTES

1 Biographical summaries are not given for living authors nor for those who are not first-named in any of the works of joint authorship.

2 The numbers refer to the Primary Literature (Section I).

Aikin, Arthur. 1773-1854 b. Warrington. Chemist and scientific writer. One of the founders of the Geological Society. [64]

Allan, Douglas A. [485, 567]

Allen, Percival [638]

Allport, Samuel. 1816-1897 b. Birmingham where he had various business enterprises; librarian to the Mason College. Pioneer in microscopic petrography. [213, 222, 228]

Anderson, Ernest Masson. 1877-1960 b. Falkirk. Outstanding for his many applications of dynamical principles to geological problems. Eminent in the work of the Geological Survey in Scotland. Murchison Medal. [440, 571, 593]

Anderson, John Graham Comrie. [556, 585, 625, 655]

Andrew, G. with Jones, [461]

Arber, Edward Alexander Newell. 1870-1918. At Cambridge from 1899. Teaching, and research on the plant remains from the Coal Measures of Britain and overseas. Made a notable excursion into the field of coastal geomorphology. [354, 384, 391, 411]

Arkell, William Joscelyn. 1904-1958 b. Highworth, Wiltshire. In a lifetime of geological research, he was, in particular, the investigator and expositor of the Jurassic system. A Senior Research Fellow of New College, Oxford, 1933-47, and of Trinity College, Cambridge, 1947-58. F.R.S. [493, 519, 532, 586, 587, 608]

Aveline, W. T. with Salter, [170]

Bailey, Edward Battersby. 1881-1965 b. Marden, in the Kentish Weald. Dedicated his life to geological research, thinking, and exposition. The geology of Scotland, particularly Dalradian and Tertiary igneous. Geological Survey, 1902-29, Professor at Glasgow, 1929-37, Director of the Geological Survey, 1937-45. Knighted, 1945. Murchison Medal : Wollaston Medal. A Royal Medal of the Royal Society. [377, 397, 412, 434, 451, 459, 498, 527, 626] with Clough and Maufe, [372]

Bakewell, Robert. 1768-1843 b. Nottingham (probably). Geologist : traveller, surveyor, instructor, and writer. [71]

Banks, Joseph. 1743-1820 The famous president of the Royal Society from 1778 to the time of his death. Baronet, 1781. [35]

Barrow, George. 1853-1932 b. London. Held high positions on the Geological Survey, 1876-1915; eastern Highlands of Scotland (metamorphic studies), North Staffordshire, Cornwall, London. Murchison Medal. [297]

Beaumont, John. d. 1731. Antiquarian, naturalist, surgeon, spiritualist. Lived at Stone Easton, Somerset. F.R.S., 1685. [9]

Bemrose, Henry Howe Arnold. 1857-1939. Printing business, public work, and geological studies concentrated on his home county, Derbyshire. Murchison Medal. [365]

Bennison, George Mills. [657]

Berger, J. F. d. 1833 Geologist; a native of Switzerland who sought refuge in Britain in 1813. [74]

Bisat, William Sawney. [452]

Blake, John Frederick. 1839-1906 b. Stoke-next-Guildford. Clergyman and scientist: mathematical, biological, geological (chiefly Jurassic and Pre-Cambrian). [217, 223, 229, 243]

Bolton, Herbert. 1863-1936 b. Bacup, Lancashire. Pre-eminent in the museum world; palaeontographer, especially of Coal Measure insects. [428, 499]

Bonney, Thomas George. 1833-1923 One of the most distinguished figures in British geology throughout the last quarter of the nineteenth century. A great teacher, a prolific writer on many subjects, geological and other, and a deep researcher in petrology. President of the British Association, 1910. Wollaston Medal. F.R.S. Honorary Canon of Manchester. [230, 231, 244, 250, 254, 285, 286] with Callaway, [238]

Borlase, William. 1695-1772 Clergyman, naturalist, antiquary. Descended from a Norman family who settled in Cornwall. [31]

Boswell, Percy George Hamnall. 1886-1960 b. Woodbridge, Suffolk. Professor (Liverpool and London) and administrator; authority on sedimentary petrology with its implications in the wider fields of pure geology and applications to industry. Tertiary of south-east England and Lower Palaeozoic of North Wales. F.R.S. [413, 441, 475, 520, 600]

Boué, Ami. 1794-1881 b. Hamburg, of Swiss parentage and French descent. Sent to Edinburgh where he graduated M.D. and took up geology. Later settled in Paris. [96]

Boulton, William Savage. 1867-1954 Assistant and disciple of Charles Lapworth; Professor (Cardiff and Birmingham). A leading official consultant on applied geology; mining, civil engineering and particularly water supply. [453, 521]

Bowerbank, James Scott. 1797-1877 Successful London business man, keen and influential palaeontologist. Studies on fossil sponges and the fossils of the London Clay, particularly the fruits. One of the founders of the Palaeontographical Society, 1847. [142]

Brewer, James. Biography unknown to the writer except that he was a medical doctor. [18]

Chapman, William. Biography unknown to the writer beyond the style 'Captain'. [32]

Charlesworth, John Kaye. [627, 634]

Chatwin, Charles P. [548, 557]

Childrey, Joshua. 1623-70 Antiquary and astrologer. Archdeacon of Sarum, 1664. [4]

Clough, Charles Thomas. 1852-1916 b. Huddersfield. Geological Survey, 1874 until his death by being run over in a railway-cutting in the Scottish coal-fields. Famous for his geological mapping, particularly in the western Highlands. Murchison Medal. [317, 372, 379]

Cobbold, Edgar Sterling. 1851-1936 b. St. Albans. Civil engineer. Moved to All Stretton, 1886, and carried out his detailed researches on the Cambrian rocks and fossils of South Shropshire. [477, 522]

Conybeare, William Daniel. 1787-1857 Geologist and divine; Dean of Llandaff, 1845. One of the band of pioneers that in the early part of the nineteenth century raised the study of rocks and fossils into the science of geology. The chief investigator and describer of the early discoveries of marine fossil reptiles in Britain. [99, 102, 103, 113] with Buckland, [112]

Cotton, G. with Hudson, [572, 577]

Cox, Arthur Hubert. 1884-1961 Chemical and industrial geologist. Professor at Cardiff. Investigator of, particularly, the Lower Palaeozoic rocks of South Wales and the Cader Idris range. [429, 460]

Craig, Gordon Younger. [651]

Craig, R. M. with Jehu, [443]

Crampton, Cecil Burleigh. 1871-1920 Geological Survey in Scotland. [403]

Crookall, R. with Welch, [547]

Currie, Ethel Dobbie. 1898-1963 b. Glasgow where she became

Assistant Curator of the Hunterian Museum. Work on Carboniferous goniatites and their significance in Scottish stratigraphy. [620]

Dale, Samuel. 1659?-1739 Physician, naturalist, antiquarian. Lived in Essex. [28]

Darwin, Charles. 1809-82 The great naturalist. [181]

Davidson, Charles Findlay. 1911-67 International authority on ore-genesis. Geology of Scotland; Professor at St Andrews, 1953. [533]

Davidson, Thomas. 1817-85 A life-work devoted to the fossil Brachiopoda. [161]

Davies, Arthur Morley. 1869-1959 b. Swansea. The doyen of British palaeontologists in his later years; teacher and writer. Imperial College of Science. Buckinghamshire became his adopted county. [398, 425, 528]

Davies, J. H. with Trueman, [483]

Davies, Kenneth Arthur. [468, 486, 523]

Davies, W. with Fearnsides, [573]

Davis, William Morris. 1850-1934 Famous American geomorphologist whose work extended to Britain on occasion. [373]

Davison, Charles. 1858-1940 Seismologist. [454]

Dawkins, William Boyd. 1837-1929 Vertebrate palaeontologist, anthropologist, economic geologist. F.R.S. Knighted, 1919. [202, 214, 245]

De la Beche, Henry Thomas. 1796-1855 After a short time in the army committed himself to a geological career; researches in Devon and Cornwall. Writer of important books on general geology. His greatest work was the founding of the Geological Survey, 1835. Knighted, 1842. Wollaston Medal. [104, 115, 116, 117, 122, 126, 130, 139, 154, 162] with Conybeare, [99]

Dearnley, Raymond. [644, 647]

Evans, John. 1823-1908 The great authority on the relics of prehistoric man. K.C.B., 1892. F.R.S. [209]

Evans, John William. 1857-1930 b. London. Scholar, philosopher, explorer; many geological researches covering a very wide range. Murchison Medal. F.R.S. [494]

Eyles, Victor Ambrose. [601]

Falcon, Norman Leslie. [642]

Farey, John. 1766-1826 b. Woburn, Bedfordshire. Estate agent to the Duke of Bedford, 1792-1802; moved to London and became a practising and consulting surveyor and geologist. [65]

Farquhar, O. C. with Read, [615]

Faujas Saint Fond, Barthelmy. 1741-1819 Born Montélimart in the Rhone valley, his family possessing the lands of Saint Fond in Dauphiné. A successful lawyer but, coming under the influence of Buffon, took up posts as a naturalist : at the Muséum d'Histoire Naturelle at Paris, later as Royal Commissioner of Mines, and later still as Professor of Geology. [48]

Fearnsides, William George. 1879-1968 b. Horbury, Yorkshire. At Cambridge, 1904-13, his earlier researches (North Wales) following the Cambridge Lower Palaeozoic tradition. Professor at Sheffield, 1913-47; consultant in the fields of coal, minerals, refractories, water supply, and engineering. Murchison Medal. F.R.S. [357, 380, 469, 524, 573]

Fitton, William Henry. 1780-1861 Physician. After marriage moved to London and devoted himself to geology. Much associated with the Geological Society. Wollaston Medal. F.R.S. [134]

Flett, John Smith. 1869-1947 Geological Survey, 1901, Director 1920-35. K.B.E., 1925. Wollaston Medal. F.R.S. [381, 392] with Dewey, [385]

Forbes, Edward. 1815-54 Botanist, zoologist, geologist; particularly in research and collections on scientific journeys and voyages. Geological Survey, 1844. In a brief career one of the most distinguished and philosophical of naturalists. F.R.S. [157, 174]

Ford, T. D. with Sylvester-Bradley, [656]

Forster, Westgarth. 1772-1835 b. Hunstanworth, County Durham. Mining surveyor. [61]

Foster, Clement le Neve. 1841-1904 Inspector of mines at home and abroad. F.R.S. Knighted, 1903. [199]

Fox-Strangways, Charles Edward. 1844-1910 b. Rewe, nr. Exeter. Geological Survey, 1867-1904. [294]

Garwood, Edmund Johnston. 1864-1949 b. Bridlington, Yorkshire. Geologist, traveller, explorer, climber, glaciologist. Great services to University College London (Professor) and the Geological Society. British geological work chiefly in investigation of Lower Carboniferous succession and, in palaeontology, of calcareous algae. F.R.S. [393, 508]

Geikie, Archibald. 1835-1924 b. Edinburgh. Pre-eminent among British geologists. The only geologist to have been President of the Royal Society (1908-12). Certainly the leading figure in British geology during the thirty years, 1882-1912. Geological Survey, 1855; Director, 1882-1901. Professor at Edinburgh, 1871-82. Twice President of the Geological Society. President of the Classical Association, 1910. Murchison Medal. Wollaston Medal. Knighted, 1891. K.C.B., 1907. O.M., 1913. Great in research and administration; above all in his writings, which are great literature on great themes. [185, 186, 190, 200, 224, 234, 239, 249, 268, 269, 288, 313, 318, 319, 346]

Geikie, James. 1839-1915 b. Edinburgh. The younger brother of Archibald Geikie. Geological Survey, 1861-82, his chief official work being on the glacial deposits and with this and independent work on other aspects of the Pleistocene he became the leading authority on that period. Professor at Edinburgh, 1882-1914. Murchison Medal. F.R.S. [215, 321, 358]

George, Thomas Neville. [478, 525, 538, 568, 636, 645] with Pringle, [560]

Gibson, Walcot. 1864-1941 Early work in Africa. Geological Survey, 1893-1925; a life-work of coalfield research. Murchison Medal. F.R.S. [341, 359, 470]

Giletti, Bruno J. [643]

Gilligan, Albert. 1874-1939 b. Wolverhampton. At Leeds, 1906; Professor there, 1922-1939. Sedimentary petrography; Yorkshire geology; geological education. [426]

Godwin-Austen, Robert Alfred Cloyne. 1808-84 Devonshire geology; south-east England; general aspects, particularly palaeo-geography; work with the Geological Society. Wollaston Medal. F.R.S. [148, 175]

Green, Alexander Henry. 1832-96 Geological Survey, 1861-74; much work chiefly on the Carboniferous of the northern Midlands and Yorkshire. Professor at Leeds, 1874-88, and at Oxford, 1888-96. F.R.S. [206, 225, 235]

Green, John Frederick Norman. 1873-1949 Colonial office. An amateur geologist of high attainments in research, with valuable work in connection with the Geological Society and Geologists' Association. [368]

Greenly, Edward. 1861-1951 Geological Survey: officially, 1889-95 (northern Highlands); unofficially, 1895-1915 (Anglesey). Lived and worked in northern Caernarvonshire, 1920 onwards. [424, 442, 500]

Greenough, George Bellas. 1778-1855 Politician, geologist, geographer. First President of the Geological Society. [91, 92]

Groom, Theodore Thomas, 1863-1942 Geologist and zoologist: Leeds, Cirencester, Reading, Birmingham, London. [326, 334, 342, 347]

Haime, J. with Milne-Edwards, [159]

Hall, James. 1761-1832 Baronet, of Dunglass, East Lothian. Friend of James Hutton. Established experimental research as a powerful aid in the investigation of geological problems. [44, 58]

Harker, Alfred. 1859-1939 b. Kingston-on-Hull. The outstand-ing figure among British petrologists of his time. St. John's College (and the Sedgwick Museum), Cambridge for 61 years. Famous for his teaching, his writings, and his research, particularly in the

Inner Hebrides. Murchison Medal. Wollaston Medal. F.R.S. [257, 276, 289, 304, 312, 343, 355, 369, 374, 422, 515]

Harkness, R. with Hicks, [207]

Harland, Walter Brian. [650, 654]

Harmer, Frederick William. 1835-1923 Public services with the city of Norwich. Geological work on the East Anglian Pliocene and Pleistocene; stratigraphy and palaeogeography; Crag Mollusca; glacial history. Murchison Medal. [322, 404]

Harrison, W. J. with Lapworth and Watts, [323]

Hatch, Frederick Henry. 1864-1932 Geological Survey as petrologist, 1886-92; mining geology, particularly in South Africa. Textbooks of mineralogy and petrology. [290, 399]

Heard, Albert. 1894-1944 University College, Cardiff. The Old Red Sandstone, petrology and palaeobotany; Lleyn peninsula, petrology and stratigraphy. [479, 566] with Matley, [503]

Heer, O. with Pengelly, [189]

Hennah, Richard. 1766-1846 Early life at St. Austell, Cornwall. Chaplain to the Plymouth Garrison, 1804-46. [84]

Henslow, John Stephens. 1796-1861 b. Rochester. Naturalist and clergyman. Professor of Mineralogy at Cambridge, 1822, and of Botany, 1827. Sedgwick's friend and Darwin's tutor. [100, 105]

Herries, R. S. with Monckton, [376]

Hickling, Henry George Albert. 1883-1954 b. Nottingham. Manchester and Newcastle-on-Tyne. Coal Measure stratigraphy and palaeontology. Murchison Medal. F.R.S. [480]

Hicks, Henry. 1837-99 b. St. David's. Physician there, 1862-71; medical and public work in London after 1871. Much geological work, particularly in Pembrokeshire. An example of an 'amateur' geologist doing work of high professional standard. F.R.S. [207, 211]

Hill, E. with Bonney, [231, 286]

Howe, J. A. with Hind, [344]

Howell, H. H. with Geikie, [185]

Huddleston, Wilfrid Huddleston. 1828-1909 (Born Simpson). Started as a lawyer, but became an 'amateur' natural scientist (and geologist in particular) of wide attainments and professional calibre. Great service to the Geological Society and the Geologists' Association. Wollaston Medal. [266] with Blake, [229]

Hudson, Robert George Spencer. 1895-1966 b. Rugby. Yorkshire became his adopted county. Leeds University (Professor, 1940-5). Iraq Petroleum Company, 1945-58. Professor at Trinity College Dublin, 1961-6. Outstanding contributions to Carboniferous stratigraphy and palaeontology in the north of England. Murchison Medal. F.R.S. [455, 572, 577]

Hughes, Thomas McKenny. 1832-1917 b. Aberystwyth. Son of the Bishop of St. Asaph. Geological Survey. Succeeded Adam Sedgwick as Professor at Cambridge in 1873 (the two holding the chair for ninety-nine years). Inspirer of the famous Cambridge school of geology, the Sedgwick Museum. F.R.S. [345]

Hutton, James. 1726-97 b. Edinburgh. The 'Founder of modern Geology'. Studied medicine, M.D. (Leyden), 1749. Returned to Edinburgh and applied himself to agriculture, eventually settling on his paternal farm in Berwickshire 1754, moving again to Edinburgh in 1768. His great geological discoveries followed. [40, 47] with Hall, [44]

Illing, Vincent Charles. 1890-1969 One of the world's leading petroleum geologists. Imperial College of Science (Royal school of Mines), 1913-55 (Professor from 1935). Murchison Medal. F.R.S. [414]

Jameson, Robert. 1774-1854 b. Leith. Studied under Werner at Freiburg and thenceforward promulgated Wernerian ideas. Professor of Natural History and Keeper of the University Museum, Edinburgh, 1804-54. [52, 57]

Jamieson, Thomas Francis. 1829-1913 b. Aberdeen. Agricultural expert. Distinguished for his researches on the glacial geology of Scotland. Murchison Medal. [179, 183, 191, 201, 216, 361]

N

Jehu, Thomas John. 1871-1943 b. Montgomeryshire. Studied science and medicine at Edinburgh, M.D. 1902. St. Andrew's, 1903; Professor at Edinburgh, 1914. Chiefly Scottish geology, but also glacial geology in Wales. [443]

Johnson, Michael Raymond Walter. [648]

Jones, Owen Thomas. 1878-1967 b. Newcastle Emlyn, Cardiganshire. The great investigator of the geology of the Lower Palaeozoic rocks of Wales. The 109 items in the bibliography of his writings—many being monographs of the first importance—cover a period of sixty-two years. Geological Survey, 1903-10. Professor at Aberystwyth, 1910-19, at Manchester, 1919-30, and at Cambridge 1930-43. Twice President of the Geological Society. Wollaston Medal. A vice-president of the Royal Society, 1940-1; a Royal Medal, 1956. [375, 394, 415, 431, 461, 558, 562, 569, 580, 589, 595, 602, 603, 628]

Jones, Thomas Rupert. 1819-1911 b. London. Began in the medical profession, but soon took up a scientific career, specialising in the micro-organisms, Foraminifera and Entomostraca, especially the fossil forms. Revision and editing of new editions of general works. Professor of Geology at Sandhurst, 1862-80. [158]

Judd, John Wesley. 1840-1916 b. Portsmouth. Work on the Jurassic and Cretaceous rocks of the East Midlands, partly independently and partly, for a few years, as a member of the Geological Survey. Independent work on the Mesozoic and Tertiary igneous rocks of western Scotland. Professor at the Royal School of Mines, 1876-1905. Fundamental petrographical studies. Wollaston Medal. F.R.S. C.B. [212, 218, 258, 263, 277, 281, 299, 300]

Jukes, Joseph Beete. 1811-69 Geological surveyor of Newfoundland, 1839-40. Naturalist with H. M. S. Fly, north-east coast of Australia, 1842-6. Geological Survey, 1846, Director for Ireland, 1850-69. Surveyor and writer. F.R.S. [167, 176]

Jukes-Browne, Alfred John. 1851-1914 Geological Survey, 1874-1902; eastern counties; Cretaceous; writer. Murchison Medal. F.R.S. [256, 270, 335]

Kellaway, Geoffrey Arthur. [617] with Hollingworth and Taylor, [574]

Kendall, Percy Fry. 1856-1936 b. Clerkenwell. Royal School of Mines. Manchester; Stockport; Leeds, 1891, Professor there, 1904-22. Yorkshire geology, particularly Coal Measures and Glacial. F.R.S. [348, 456]

Kennedy, William Quarrier. [581, 596, 604, 610, 629, 637]

Kent, Percy Edward. [658] with Falcon, [642]

Kidston, Robert. 1852-1924 Carboniferous flora and the early fossil plants of the Middle Devonian at Rhynie, Aberdeenshire. Murchison Medal. F.R.S. [306, 419, 436]

King, William Bernard Robinson. 1889-1963 A Yorkshireman. Geological Survey. Water-supply in France during the 1914-18 war. Cambridge, 1920-31. Professor at University College London, 1931-9. Geological adviser in France, 1939-43. Professor at Cambridge, 1943-54. Murchison Medal. F.R.S. [621]

King, William Wickham. 1862-1959 One of the last of the self-taught amateur geologists. Research on the then wilderness of the Shropshire-Worcester-Hereford Old Red Sandstone. [530]

Kitchin, E. L. with Lamplugh and Pringle, [444]

Lake, Philip. 1865-1949 Work chiefly at, and from, Cambridge; head of geography, 1908-27. Geological work in North Wales; palaeontology of trilobites. Writer. [362, 382]

Lambert, R. St J. with Giletti and Moorbath, [643]

Lamplugh, George William. 1859-1926 A Yorkshireman. Amateur turned professional. Geological Survey, 1892-1920. Yorkshire geology, particularly Cretaceous and glacial. Isle of Man. South-east England (administration and research). Wollaston Medal. F.R.S. [278, 291, 315, 350, 427, 444]

Lang, William Dickson. 1878-1966 Geological Department of the British Museum (Natural History), 1902-38 (Keeper from 1928). Corals and Polyzoa were his special groups of fossils. Holidays on the Dorset coast resulted in his detailed studies of the Lias there. F.R.S. [405, 445, 471, 487]

Lang, W.H. with Kidston, [419]

Lankester, E. R. with Traquair and Powrie, [205]

Lapworth, Charles. 1842-1920 b. Faringdon, Berkshire. Famous for his work on graptolites and the Lower Palaeozoic rocks. In 1864 became a schoolmaster in Scotland, Galashiels and St. Andrews, and meanwhile unravelled the structure of key areas of the Southern Uplands. Professor at Birmingham, 1881. Midlands geology and the North-west Highlands. Founded the Ordovician system. Wollaston Medal. A Royal medal of the Royal Society. [236, 240, 241, 250, 253, 259, 271, 292, 307, 323]

Lee, Gabriel Wharton. 1880-1928 Palaeontologist and surveyor on the Geological Survey of Scotland; particularly Carboniferous palaeontology and Mesozoic stratigraphy and palaeontology. [516] with Bailey, [459]

Lees, George Martin. 1898-1955 Carried out and inspired the geological exploration of the Middle East in his capacity of Chief Geologist to the Anglo-Iranian Oil Company, 1932-53. Initiated a new attempt to discover oil in England in 1933. F.R.S. [582]

Leigh, Charles. 1662-?1701 A Lancashire physician and naturalist. F.R.S. [19]

Leitch, Duncan. 1904-56 b. Glasgow. Upper Carboniferous stratigraphy and palaeontology. Glasgow University, 1938. Professor at Swansea, 1947-56. [570] with Weir, [554]

Leland, John. 1506?-52 Antiquary and topographer. [1]

Lewis, Henry Carvill. 1853-88 b. Philadelphia, U.S.A. Professor, 1880-8, and Pennsylvanian Geological Survey; research on glacial geology. In Europe, 1885-8, studying the glacial geology of the British Isles. [308]

Lewis, Herbert Price. 1895-47 Sheffield University, 1921-31, Professor at Aberystwyth, 1931-47 Stratigraphy, petrology, and palaeontology, specialising in Carboniferous corals and various micro-fossils. [501]

Lhwyd, Edward. 1660-1709 b. Oswestry. Celtic scholar and naturalist. Keeper of the Ashmolean Museum, Oxford, 1690-1709. [15, 17]

Lister, Martin. 1638-1712 b. Radclive, Buckinghamshire. Antiquarian, conchologist, and general naturalist. Ancestrally associated with the Craven district of Yorkshire. Physician at York. [7, 11, 12, 14]

Lonsdale, W. with Sedgwick and Murchison, [144]

Lyell, Charles. 1795-1875 b. Kinnordy, Angus, Scotland. Established and interpreted the principles of geology. A great traveller. Much concerned with the Geological Society : twice President; Wollaston Medal. A Royal Medal and the Copley Medal of the Royal Society. [125, 138, 192]

M'Coy, F. with Sedgwick, [173]

Macculloch, John. 1773-1835 b. Guernsey. Geologist and physician. The early surveyor of the geology of Scotland. F.R.S. [85, 93]

MacGregor, A. G. with Macgregor, [550]

Macgregor, Murray. 1884-1966 b. Glasgow. Became Scotland's most eminent coalfield geologist. Geological Survey, 1909-45. [462, 502, 550]

McIntyre, Donald B. [622]

Maclaren, Charles. 1782-1866 Established and edited the 'Scotsman'; other literary activities including geological. [140]

McLintock, William Francis Porter. 1887-1960 Geological Survey, 1907-50; Director, 1945-50; establishment of the new museum in South Kensington. C.B. [597]

Mantell, Gideon Algernon. 1790-1852 b. Lewes. Surgeon, naturalist, geologist, palaeontologist, lecturer, collector, writer. Lived in Sussex till 1839 when he moved to London. F.R.S. [106, 114]

Marr, John Edward. 1857-1933 Born at Morecombe, of Scottish descent. Studied Lower Palaeozoic of Bohemia and Scandinavia. Cambridge, 1881-1933; Professor 1917-33. The geology of the Lower Palaeozoic rocks of Wales and, particularly, the English

Morris, T. O. with Fearnsides, [469]

Morton, George Highfield. 1826-1900 b. Liverpool. A business man devoting all his leisure time to geological research. Very influential in the geological life of Liverpool. [193]

Morton, John. 1671?-1726 Rector of Great Oxenden, North-amptonshire. Naturalist. F.R.S. [22]

Murchison, Roderick Impey. 1792-1871 b. Tarradale, Ross-shire. After the army and dallying with art, became an enthusiastic and famous geologist. Founded the Silurian, Devonian, and Permian systems after great work in the Welsh Borderland, South-west England, and Russia. Director of the Geological Survey, 1855-71. Twice President of the Geological Society; President of the British Association, 1846; President of the Royal Geographical Society. Wollaston Medal. F.R.S. Knight, 1846. Baronet, 1866. [118, 120, 141, 169] with Sedgwick and Lonsdale, [144]

Neaverson, Ernest. [489]

Necker, Louis Albert. 1786-1861 b. Geneva. Edinburgh University, 1806, travelling in Scotland and England. Professor of Natural History at Geneva, 1810-c1835, paying visits to Scotland and finally settling in Skye. [60]

Newton, Edwin Tulley. 1840-1930 Palaeontologist to the Geological Survey, 1865-1905. Renowned for the delicate dissection and minute description of certain vertebrate fossils. F.R.S. [237, 273, 301]

Nicholas, Tressilian Charles. [408]

Nicholson, H. A. with Marr, [272, 293]

Nicol, James. 1810-79 b. Inverleithen, Peebles-shire. University studies (Scotland and Germany), assistant to the Geological Society, Professor of Natural History at Aberdeen, 1853-79. Researches on the geology of the Southern Uplands and Northern Highlands of Scotland. [177, 187]

Oakley, K. P. with Edmunds, [549]

Otley, Jonathan. 1766-1856 b. Grasmere. Topographical writer on the Lake District, and its earliest geologist. Moved to Keswick in 1791 and lived there for the rest of his long life. [110]

Owen, Edward. Biography unknown to the writer. [30]

Owen, George. 1552-1613 b. Henllys, Pembrokeshire. Vice-admiral of Pembroke and Cardigan. Historian of Pembrokeshire. [3]

Owen, Richard. 1804-92 Surgeon, anatomist, famous vertebrate palaeontologist. Wollaston Medal. F.R.S. K.C.B. [163]

Owen, T. R with Anderson, [655]

Packe, Christopher. 1686-1749 b. St. Albans. Physician. Settled at Canterbury. [29]

Parkinson, James. 1755-1824 Born at Hoxton (London), where he lived all his life. Physician and surgeon, geologist, social reformer. One of the founders of the Geological Society. [55, 68, 107]

Peach, Benjamin Neave. 1842-1926 b. Cornwall. In the forefront of Scottish geology for half a century. His name is inseparably linked with that of John Horne. Geological Survey, 1862 onwards. Murchison Medal and Wollaston Medal (both with Horne). F.R.S. [242, 246, 248, 295, 327, 366, 504]

Pengelly, William. 1812-94. The energetic and highly successful educationist, scientist and geologist (in all its aspects) of Torquay. F.R.S. [171, 189]

Phemister, James. [552]

Phillips, Frank Coles. [559, 578, 611]

Phillips, John. 1800-74 b. Marden, Wiltshire. Of Welsh paternal ancestry; nephew and pupil of William Smith. Keeper of the York Museum, 1825-40. Successively Professor at London, Dublin, and Oxford. Keeper of the Ashmolean Museum, 1854-70. Wollaston Medal. F.R.S. Wide and deep geological research; writer of comprehensive and 'classical' works on geology. [123, 129, 135, 137, 146, 155, 172, 208, 261]

Phillips, William. 1773-1828 Printer and bookseller; mineralogist and geologist; one of the founders of the Geological Society. F.R.S. [78, 90, 94] with Conybeare, [103]

Playfair, John. 1748-1819 b. Benvie, near Dundee. Mathematician and geologist. Professor of mathematics, and later of natural philosophy, at Edinburgh. F.R.S. [54]

Plot, Robert. 1640-96 b. Borden, Kent. Antiquary and naturalist. Secretary and editor to the Royal Society. Much associated with Oxford; first Keeper of the Ashmolean Museum; Professor of Chemistry. [10, 13]

Pocock, Roy Woodhouse. [539] with Cobbold, [522]

Powrie, J. with Traquair and Lankester, [205]

Prestwich, Joseph. 1812-96 While engaged in the family wine-merchant business travelled extensively and undertook much geological investigation. Professor at Oxford, 1874-88. Wollaston Medal. A Royal Medal of the Royal Society. Knighted, 1896. [143, 160, 164, 196, 264, 296]

Pringle, John. 1877-1948 b. Selkirk. A palaeontologist on the Geological Survey, 1901-37. Unofficial work in Wales and Scotland. [540, 560] with Davies, [398] with Lamplugh and Kitchen, [444] with Lee, [516]

Pryce, William. 1725?-90. b. Redruth, Cornwall. Surgeon and antiquary. Mine-owner and authority on mine-working. [36]

Pugh, William John. [446, 490, 495] with Jones, [415, 580, 595, 603]

Ramsay, Andrew Crombie. 1814-91 Geological Survey, 1841, Director 1871-81. Professor at University College London, 1848. Wollaston Medal. A Royal Medal of the Royal Society. Knighted, 1881. [147, 152, 166, 182, 184, 194, 195]

Rastall, Robert Heron. 1871-1950 Associated with Cambridge, 1899-1942; retired to the family estate near Whitby. [383] with Lake, [382] with Hatch, [399]

Rayner, Dorothy Helen. [653]

Read, Herbert Harold. 1890-1970 Trained under Watts at Imperial College of Science. Geological Survey, 1914-31. Famous researches, and expositions in presidential addresses, on the stratigraphy, structure, metamorphism and associated igneous geology, of the Dalradian of Scotland (and, later, in Ireland). Professor at Liverpool, 1931; at Imperial College, 1939. President of the International Geological Congress, London, 1948. Wollaston Medal. A Royal Medal of the Royal Society. [447, 448, 464, 481, 510, 531, 541, 542, 553, 614, 615, 631]

Reed, Frederick Richard Cowper. Most of his life spent in Cambridge; palaeontologist and curator at the Sedgwick Museum. Writer of palaeontographical monographs, particularly on Scottish and Indian fossils. Travel overseas; writer on the geology of the British Empire. [351, 420]

Reid, Clement. Geological Survey, 1874-1913. His most important work was on the plant-life and conditions of the Pliocene and Pleistocene periods. F.R.S. [282, 401]

Reynolds, Sidney Hugh. 1867-1949 Associated with Bristol University, 1894 onwards (Professor, 1910). Lower Palaeozoic of Ireland and Wales but particularly Lower Carboniferous of the Bristol district. Geological photography. [472]

Richardson, Linsdall. 1881-1967 A life-time of geological work based on Bristol, Cheltenham, and Birmingham. [367, 417]

Richardson, W. A. with Lang and Spath, [445]

Richey, James Ernest. 1886-1968. b. Desertcreat Rectory, County Tyrone. Trinity College Dublin. The leading authority on Tertiary volcanic complexes in Scotland and Ireland. Geological Survey, 1911-46. [505, 518, 543]

Roberts, T. with Marr, [260]

Robinson, Thomas. d. 1719. Rector of Ousby, Cumberland. [21]

Rowe, Arthur Walton. 1858-1926 Medical man of Margate and highly skilled geological and palaeontological investigator of the Chalk. [328, 336]

Rutley, Frank. 1842-1904 Mineralogist and one of the early practitioners of microscopic petrography. Geological Survey, 1867-82. Lecturer at the Royal School of Mines, 1882-98. [262]

Salter, John William. 1820-69 Worked with J. de C. Sowerby, and with Sedgwick at Cambridge. Geological Survey as palaeontologist 1846-63. His researches were particularly on the trilobites. [170, 197]

Satterthwaite, G. E. with Kent and Spencer, [658]

Sedgwick, Adam. 1785-1873 b. Dent, Yorkshire. His main home from 1810 was Trinity College, Cambridge. Professor of Geology, 1818-73. Began in 1831 the investigations which founded the Cambrian system. Devonian system with Murchison. North of England researches. College business and general procrastination delayed publications, but his personality and advocacy greatly raised the status of geology. First President of the British Association, 1833. Ordained priest, 1818; Canon of Norwich, 1834. Wollaston Medal. Copley Medal of the Royal Society. [124, 127, 131, 132, 136, 144, 149, 153, 168, 173]

Seeley, H. G. with Phillips and Etheridge, [261]

Seward, Albert Charles. 1863-1941 Botanist and geologist; eminent palaeobotanist. Professor of botany, Cambridge, 1906-36. Master of Downing College. Murchison Medal. Wollaston Medal. F.R.S. Knighted, 1936. [310, 324, 337, 511]

Sherborn, C. D. with Woodward, [284]

Sherlock, Robert Lionel. 1875-1948 Geological Survey, 1903-37. The authority on the Permian and Triassic systems at home and abroad. [388, 432, 473, 544]

Sibly, Thomas Franklin. 1883-1948 b. Bristol. King's College, London; Professor at Cardiff, 1913-18, at Newcastle, 1918-20. Carboniferous Limestone. High academic and administrative posts. Knighted 1938. K.B.E., 1943. [370, 423]

Sinclar, George. d. 1696 Probably a native of East Lothian. Professor of philosophy and mathematics at Glasgow. Meteorologist, hydrologist, and general scientist and inventor. [8]

History). Specialist on ammonites. F.R.S. [449] with Lang and Richardson, [445] with Lang, [471] with Lang and others, [487]

Spencer, A. M. with Kent and Satterthwaite, [658]

Stamp, Laurence Dudley. 1898-1966 Academic distinction and public service in geography; research and writing in geology. Knighted, 1965. [433]

Steers, James Alfred. [584]

Stewart, F. H. with Johnson, [648]

Strachey, John. 1671-1743 Of Sutton Court, Chew Magna, near Bristol. Geology, particularly in connection with coal mines, was his chief interest. [23]

Strahan, Aubrey. 1852-1928 Geological Survey 1875-1920; Director from 1914. Wollaston Medal. F.R.S. K.B.E., 1919. [325, 329]

Strange, John 1732-99 Diplomatist, naturalist, antiquary. [39]

Straw, Stephen Henry. 1891-1963 After Nottingham, University of Manchester, 1923-58. Research on the Ludlovian division of the Silurian. [561]

Stubblefield, Cyril James. [482, 652] with Evans, [494] with Dunham, [576] with Edwards, [594]

Stukeley, William. 1687-1765 b. Holbeach, Lincolnshire. A medical man; entered the Church. Antiquary, naturalist, general virtuoso. [24]

Sutton, John. [612, 618, 623, 632, 640]

Swinnerton, Henry Hurd. 1875-1966 The 'natural sciences', particularly geology and palaeontology, and general affairs, at Nottingham from 1902 onwards. [450, 546]

Sylvester-Bradley, Peter Colley. [656]

Taitt, A. H. with Lees, [582]

Tate, R. with Blake, [223]

Taylor, James Haward. 1909-68 Geological Survey, 1935-49; Professor at King's College, London. Specially noteworthy work in the Midlands. [605, 649] with Hollingworth and Kellaway, [574] with Hollingworth, [580, 609] with Kellaway, [617]

Teall, Jethro Justinian Harris. 1849-1924 Petrographer to the Geological Survey, 1888; Director, 1901-14. Wollaston Medal. F.R.S. Knighted, 1916. [274, 364]

Thomas, Herbert Henry. 1876-1935 Geological Survey, 1901-35. Chiefly petrography. Murchison Medal. F.R.S. [390]

Tilley, Cecil Edgar. [457, 465]

Topley, William. 1841-94. Geological Survey, 1862-94. F.R.S. [219] with Foster, [199]

Townsend, Joseph. 1739-1816 Physician. Rector of Pewsey, Wiltshire. One of the original honorary members of the Geological Society. [72]

Traquair, Ramsay Heatley. 1840-1912 The great investigator of the fossil fishes of the Palaeozoic formations, particularly those of Scotland. After demonstrating in human anatomy at Edinburgh and professing Natural History (Cirencester) and Zoology (Dublin), he was Keeper of the Natural History collections at Edinburgh, 1873-1906. A Royal Medal of the Royal Society. [205, 330, 352]

Trechmann, Charles Taylor. 1885-1964 Geologist, chemist, archaeologist, traveller. In British geology (of which he was a notable character) contributions to the geology of his native County Durham. [402, 406, 409, 466]

Trotter, Frederick Murray. 1897-1968 Geological Survey, 1921-63. North of England and South Wales, Carboniferous and Pleistocene, economic applications. Murchison Medal. [496]

Trueman, Arthur Elijah. 1894-1956 b. Nottingham. Researcher, teacher, educationist, administrator, writer. Palaeontologist and Coal Measure geologist. Professor at Swansea (1920), Bristol

(1933), and Glasgow (1937). Wollaston Medal, 1955. F.R.S. K.B.E., 1951. [439, 483, 526, 564, 591, 624] with Dix, [507]

Tyrrell, George Walter. 1883-1961 b. Hanwell, Middlesex. One of the most distinguished of world petrologists. In Britain his chief work was in Scotland. For long associated with Glasgow University. [421, 474, 491]

Ussher, William Augustus Edward. 1849-1920 Geological Survey, 1868-1909. The South-western counties of England. Murchison Medal. [283, 353]

Vaughan, Arthur. 1868-1915. b. London. Initiated a new era in the study of the Lower Carboniferous by establishing its zonal stratigraphy. Clifton, 1891-1910. Lecturer in geology at Oxford, 1910-15. [360, 410] with Dixon, [386]

Walcott, John. Biography unknown to the writer. [38]

Ward, James Clifton. 1843-80. Geological Survey, 1865-78, his chief work being done in the Lake District. Scientific activity in Cumberland; curate at St. John's, Keswick, 1878. [220, 221, 226]

Watson, Janet. with Sutton, 612, 618, 623, 632, 640]

Watson, White. 1760-1835. Sculptor, marble-worker, and mineral-dealer. Lived at Bakewell, Derbyshire, where travellers visited his museum-shop. [69, 73]

Watts, William Whitehead. 1860-1947 b. Broseley, Shropshire. 'For sixty years the teacher, trusted counsellor and friend of successive generations of English geologists.' Cambridge 'Extension' lecturer, 1881-91, Deputy Professor at Birmingham, 1883, and at Oxford, 1888. Cambridge, 1888-91. Geological Survey 1891-7. Assistant Professor of Geology, Birmingham, 1897; Professor of Geography, Birmingham, 1904-6; Professor of Geology, Imperial College of Science, 1906-30. Murchison Medal. Wollaston Medal. F.R.S. In particular, the investigator and exponent of the geology of South Shropshire and Charnwood Forest. [467, 592] with Lapworth, [307] with Lapworth and Harrison, [323]

Webster, Thomas. 1773-1844 b. Orkneys. Educated Aberdeen; came to London as an architect. Taking up geology, was the chief professional member of the Geological Society staff. Lecturer. First Professor of Geology at University College London, 1841. [75, 80]

Weir, John. [554]

Welch, Francis Brian Awburn. [547]

Wells, Alfred Kingsley. [565] with Cox, [429]

Whidborne, George Ferris. 1846-1910 Descendant of a well-known Devon family. Clergyman and palaeontologist. Collector and describer of fossils of the Devonian system. [279]

Whitaker, William. 1836-1925 b. London. Geological Survey, 1857-96. Very extensive work in the south-east of England. Became a consultant on water-supply problems. Bibliographer of geology. Gave great assistance to local scientific societies. [204, 280, 331]

Whitehead, T. H. with Pocock, [539]

Whitehurst, John. 1713-88 b. Congleton, Cheshire. Clock-maker at Derby (town hall and church clocks). Settled in London, being appointed 'Stamper of the Money-weights'. A leading authority in the mechanical and natural sciences. F.R.S. [37]

Whittard, Walter Frederick. 1902-66 Imperial College of Science. Professor at Bristol, 1937-66. Lower Palaeozoic stratigraphy and palaeontology, particularly in Shropshire. Geology of the English Channel. Murchison Medal. F.R.S. [492, 616, 630]

Wilcock, B. with Harland and Smith, [650]

Williams, Alwyn. [619, 641, 646]

Williams, Howel. [484, 512]

Williams, John. Biography unknown to the writer except that he was a 'director of mines'. [43]

Wills, Leonard Johnston. [438, 458, 497, 598]

Wilson, T. S. with Matley, [583]

Winch, Nathaniel John. 1769?-1838 Botanist and general naturalist. Researches in the north of England. [88]

Wood, Alan. [658]

Wood, Ethel Mary Reader. (Dame Ethel Shakespear) 1871-1945 Newnham College, Cambridge. Research Assistant to Lapworth at Birmingham, 1896-1906. Graptolites and stratigraphical graptolite-faunas. Eminent public services. [338] with Elles, [340]

Wood, Searles Valentine (the elder). 1798-1880 The great collector, student, and describer of the Tertiary Mollusca, particularly those from the Pliocene of his native East Anglia. Wollaston Medal. [156]

Wood, Searles Valentine (the younger). 1830-84 Son of S. V. Wood, the elder, carrying on the study of the geology of East Anglia, especially the glacial deposits. For some time a solicitor at Woodbridge. [247]

Woods, Henry. 1868-1952 Palaeontologist, lecturer, unofficial kindly adviser, and geological librarian at Cambridge, 1892-1950. Wollaston Medal. F.R.S. [302, 332, 396]

Woodward, Arthur Smith. 1864-1944 British Museum (Natural History), 1882-1924. Great authority on fossil fishes. F.R.S. Knighted, 1924. [284, 349, 418]

Woodward, Horace Bolingbroke. 1848-1914 Geological Survey, 1867-1908. Jurassic, water-supply, history of geology. Notable exponent of the art of scientific literature. Murchison Medal. Wollaston Medal. F.R.S. [227, 303]

Woodward, John. 1665-1728 b. Derbyshire. Physician and geologist. Founder of the Woodwardian lectureship (later, professorship) at Cambridge. F.R.S. [16, 26, 27]

Wooller, J. with Chapman, [32]

Wray, Disney Alexander. [555]

Wright, A. E. with Bennison, [657]

Wright, James. 1878-1957 In business at Kirkcaldy, Fife. A

O

world authority (as an amateur) on the Crinoidea, particularly those from the Carboniferous. [607]

Wright, Thomas. 1809-84 b. Paisley. Physician and surgeon at Cheltenham where, in addition to his professional duties, he carried out great palaeontographical work, particularly the description of the Mesozoic Echinodermata. Wollaston Medal. F.R.S. [178, 198]

Wright, William Bourke. 1876-1939 b. Dublin. Geological Survey (Ireland, England and Scotland), 1901-39. Lancashire coalfield and, particularly, glacial geology. [407]

Wroot, H. E. with Kendall, [456]

Young George. 1777-1848 b. near Edinburgh. Theologian, topographer, geologist. Pastor at Whitby, 1806-48. [108]

Appendix C

Index of Names: places, counties, stratigraphical divisions, fossils

NOTE

In each item the first set of numbers refers to titles of the Primary Literature, the second set, in italics, to the Themes